AA003101

Separations Division

Core Programming Topic at the 2011 AIChE Spring Meeting and 7th Global Congress on Process Safety

Chicago, Illinois, USA
13 – 17 March 2011

ISBN: 978-1-61782-898-0

Printed from e-media with permission by:

Curran Associates, Inc.
57 Morehouse Lane
Red Hook, NY 12571

Some format issues inherent in the e-media version may also appear in this print version.

Printed by Curran Associates, Inc. (2011)

For permission requests, please contact AIChE
at the address below.

AIChE
3 Park Avenue
New York, NY 10016-5991

Phone: (203) 702-7660
Fax: (203) 775-5177

www.aiche.org

Additional copies of this publication are available from:

Curran Associates, Inc.
57 Morehouse Lane
Red Hook, NY 12571 USA
Phone: 845-758-0400
Fax: 845-758-2634
Email: curran@proceedings.com
Web: www.proceedings.com

TABLE OF CONTENTS

(8a) Simulated Moving Beds-Principles and Design Tools

Nien-Hwa Linda Wang, Purdue University, West Lafayette, IN

Batch adsorption and chromatography processes are highly selective separation techniques. They have been widely used for the analysis of complex mixtures and purification of chemicals, biochemicals, and pharmaceuticals. A continuous adsorption (or "Simulated Moving Bed") process can have significantly higher yield, higher purity, higher throughput, and lower solvent consumption than batch chromatography. SMB processes have been used for the production of hydrocarbons since 1970's, high-fructose corn syrup since 1980's, and chiral pharmaceuticals since 1990's. SMB, however, has not been widely used for multi-component separation. A major barrier is the lack of efficient methods and tools to design SMB that can recover a product with high purity and high yield from a mixture of three or more components. A second barrier is the lack of SMB equipment for multi-component separation. This presentation will introduce the fundamental principles of SMB, the methods and tools developed at Purdue to overcome the major barriers, and the applications of the tools in developing a tandem SMB for insulin purification from a ternary mixture.

See more of this Session: YP Tutorial: Separations
See more of this Group/Topical: Separations Division - Please See Topical 8

(8b) Gas Separations Using Polymer Membranes

Benny D. Freeman, Chemical Engineering, University of Texas at Austin, Austin, TX

This presentation will discuss structural features important in the use of polymers as rate-controlling membranes for gas separations. In particular, materials having desirable combinations of high permeability and high selectivity based upon solubility selectivity (e.g., butane removal from natural gas, CO_2 separation from H_2 or N_2) or diffusivity selectivity (e.g., CO_2 removal from natural gas) will be presented. For example, cross-linked poly(ethylene oxide) (XLPEO) polymers, which are flexible, rubbery polymers identified as promising materials to remove polar and acid gases, such as CO_2, from mixtures with light gases, such as H_2. One member of this family of materials was reported to have a CO_2 permeability coefficient of approximately 500 Barrer and a CO_2/H_2 mixed gas selectivity of 30 at -20C. Such materials achieve high selectivity based upon their high solubility selectivity favoring CO_2 transport.

Polymers can also be tailored to achieve high selectivity based upon high diffusivity selectivity. In this case, highly rigid, glassy polymers with proper free volume element size and size distribution are desirable. Polyimides with ortho-position functional groups may be solution-processed to form conventional films and membranes. Such materials can undergo thermal rearrangement to form highly rigid benzoxazole or benzithiazole structures having very high permeability coefficients and high selectivity. For example, one member of this family was prepared having a CO_2 permeability coefficient of 1610 Barrer and a CO_2/CH_4 selectivity, under mixed gas conditions, of 42-46, depending on the partial pressure of CO_2 in the mixture. These thermally rearranged (TR) polymers are insoluble in common solvents, giving them good chemical stability, and highly thermally stable, which are important attributes for membranes that would be used in chemically or thermally aggressive environments.

See more of this Session: YP Tutorial: Separations
See more of this Group/Topical: Separations Division - Please See Topical 8

(8c) Functionalized Membranes for Environmental Applications: An Overview

D. Bhattacharyya, Chemical and Materials Engineering, University of Kentucky, Lexington, KY

Membrane processes such as, Reverse Osmosis, Nanofiltration, Ultrafiltration, and Microfiltration have provided many successful applications ranging from high quality water production to material recovery, but incorporation of biological aspects in membranes should add immense value in the area of separations and material capture. The questions are: can we make membranes which can capture toxic metals at high capacity? Can we add functional moiety in pores which can go through conformation changes and results in tunable separations? Can we make layer-by-assembly in pores to immobilize enzymes for catalysis? The answers to all of these are yes if we can integrate life sciences field with synthetic membranes to make tunable and functionalized materials. This overview presentation will cover some tunable aspects of separations and green synthesis of water related chemicals.

See more of this Session: YP Tutorial: Separations
See more of this Group/Topical: Separations Division - Please See Topical 8

Hydrogen Separation with "Pore-fill" Type Palladium Membrane

Tianze Zhao[1,2], Junya Okazaki[1], David A. Pacheco Tanaka[1,3],
Toshishige M. Suzuki[1], and Yoshito Wakui[1]

[1] *Research Center for Compact Chemical System, National Institute of Advanced Industrial Science and Technology (AIST), 4-2-1, Nigatake, Miyagino-ku, Sendai 985-8551, Japan*

[2] *Division of Engineering, Graduate School of Tohoku Gakuin University, 13-1, Chuou-1chome, Tagajo 985-8537, Japan*

[3] *Faculty of Engineering, University of Porto, Rua Dr. Roberto Frias, s/n, Porto 4200-465, Portugal*

Abstract

The mechanical durability and the thermal stability of thin palladium membrane must be improved for the practical using in hydrogen recovery from mixed gas. Since conventional thin palladium membrane should be easily damaged during handling, we have recently developed a unique fabricated "pore-fill" type membrane where palladium particles fill the nano-sized pores of a supporting matrix. This pore-fill configuration could improve the mechanical durability of the membrane during handling since palladium layer is not exposed directly on the top surface. Moreover, nano-size palladium particles could suppress the hydrogen embrittlement caused by the phase transition of palladium hydride due to the decrease in internal stress associated with α–β phase transition. In this study, we optimized the electroless plating conditions for the improvement of the hydrogen perm-selectivity of pore-fill type membrane. The hydrogen/nitrogen selectivity increased by addition of a surfactant, Triton® X-100, to the plating solution, and reached around 20000 at 600°C when 0.05 wt.% of Triton® X-100 was added. The hydrogen permeance reached 3.42 μmol m^{-2} s^{-1} Pa^{-1} at the same experimental conditions. The excellent selectivity could be attributed to the gas tightness of the pore-fill type membrane. After 100 hours of gas permeation test at 600°C, nitrogen leak did not increase. This indicates that the pore-fill type palladium membrane can be applicable to hydrogen separation at elevated temperatures.

Key wards: Pore-fill type membrane, Palladium membrane, Hydrogen separation, Thermal stability

1. Introduction

Since the remarkable permeability and exclusive selectivity for hydrogen, palladium membrane has been regarded as a promising material for the continuous separation and purification of hydrogen [1]. In order to attain high hydrogen permeation flux and minimize the material costs, efforts have been made to fabricate ultrathin palladium membrane [1,2]. However, the significant drawback of thin membranes is their limited durability. They are easily damaged during handling and by thermal process resulting in peeling and generation of cracks and pinholes [3]. Conversely, "pore-fill" type membrane where palladium particles fill the pores of a supporting matrix is an attractive alternative against the conventional thin film membranes [4,5]. Figure 1 shows the feature of a pore-fill type palladium membrane.

Figure 1. Configuration of the "pore-fill" type palladium membrane.

This configuration could provide easier handling of the membrane since palladium layer is not exposed directly on the top surface. In addition, the nano-size palladium particles could suppress the hydrogen embrittlement caused by the phase transition of palladium hydride [4,6]. Due to a large surface area of palladium nanoparticles, hydrogen loading preferentially occurs on the surface rather than the interior of particle. This nano-size effect substantially suppresses the internal stress associated with $\alpha-\beta$ phase transition of palladium hydride [6-8]. As previously reported, we have demonstrated the better tolerance of pore-fill type palladium membrane during hydrogen and nitrogen permeation tests below the critical temperature ($< 300°C$) [4,5].

For the practical applications of pore-fill type palladium membrane at elevated temperatures, the thermal stability of the membrane must be improved. Yttria-stabilized zirconia (YSZ) is attributed to a desirable material for the membrane substrate because of its thermally stable phase structure and chemical inertness in both oxidative and reductive environments [9,10]. Besides, the thermal expansion coefficient of zirconia is comparable to that of palladium [11]. Therefore, the use of YSZ ensures better thermal and mechanical properties of the palladium membrane systems.

The major challenge for pore-fill type membrane is improvement of the hydrogen selectivity. Thus, palladium particles are electrolessly deposited in the nano-size pores of the supporting materials. When the nano-size pores are not sufficiently filled, the gas tightness of the membrane cannot be ensured. In this study, we investigated the effect of a surfactant, Triton® X-100, to the electroless plating solution for the improvement of the hydrogen selectivity of pore-fill type membranes. The obtained membranes were characterized by SEM-EDX analysis. In addition, hydrogen or nitrogen gas was supplied to the membranes to evaluate the changes in hydrogen permeability and gas selectivity at temperatures of 300–600°C.

2. Experimental

2.1. Fabrication of YSZ tube

A homemade porous YSZ tube was used as the membrane support. The tube was fabricated by a dry-wet spinning method described in Ref. 12. A suspension including YSZ powder slurry (8 mol% Y_2O_3, TZ-8YS; Tosoh Corp.), polymer, and organic solvent was extruded to form a green YSZ tube. Then the green tube was calcined at 1350°C for 1 h, to obtain a symmetric porous tube (o.d., 2.1 mm; i.d., 1.5 mm).

2.2. Membrane fabrication

The surface of both YSZ tube ends were coated with enameled glass, leaving 5 cm of uncoated part where a pore-fill type membrane was fabricated. Then it was dipped in the YSZ sol containing 0.2 wt.% of YSZ powder. After the calcination at 500°C for 1 h, a thin porous YSZ layer was formed on the top surface of the tube. The modified tube was activated with palladium nuclei by dipping into 0.6 wt.% chloroform solution of palladium acetate, followed by reduction with 2 M hydrazine solution. By evaporation of chloroform, palladium acetate was carried to the top surface of the tube in concentrated form. Then palladium was reduced by dipping the tube into aqueous hydrazine solution for 15 min at room temperature. This cycle was repeated eight times.

Since the excess palladium nuclei on the top surface of the tube was expected to prevent the further layering omit; the thin YSZ layer was removed by ultrasonic agitation. Then the tube was again dipped in the YSZ sol for coating of the top surface. This new YSZ layer enabled the protection of the pore-fill layer where palladium particles fill the nano-sized pores of the supporting matrix. Then the activated tube was immersed in the electroless plating solution at 50°C. The plating solution was composed of 0.01 M palladium chloride, 5 M ammonia 0.12 M Na_2EDTA and, 0.13 M hydrazine as a reducing agent. A surfactant, Triton® X-100 (Wako), was also added. To facilitate the transportation of the plating solution into the pores of the YSZ tube, the solution was circulated from external to internal surface using a peristaltic pump.

2.3. Gas permeation test

The permeability and the selectivity for hydrogen gas of the membrane were estimated in the equipment depicted in Figure 2. A membrane specimen was fixed in a stainless steel tube and placed in a cylindrical furnace. Hydrogen or nitrogen gas was introduced on the membrane from the outer shell, and the amount of permeated gas was measured using a soap-bubble flow meter at 300–600°C with transmembrane pressure differences of 10–200 kPa. At this time, the inner pressure of the composite tube was kept at atmospheric pressure. The ideal separation factor was determined based on the ratio of the hydrogen to nitrogen permeation flux at the same temperature and the transmembrane pressure difference.

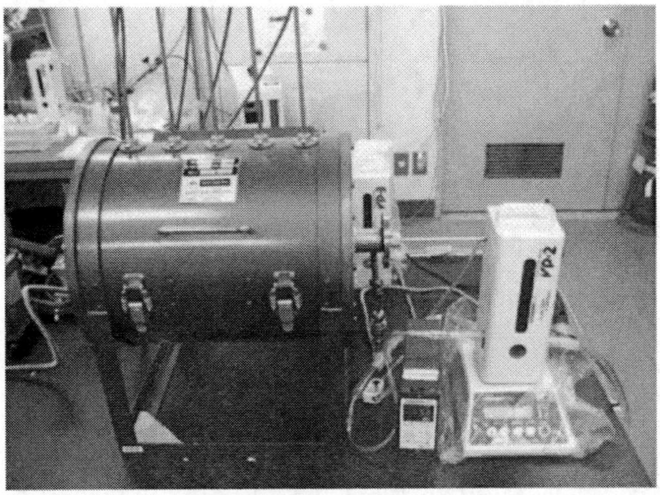

Figure 2. Equipment used for gas permeation test of pore-fill type membranes.

2.4. Characterization

A scanning electron microscope (SEM, Hitachi S-800) equipped with an energy dispersive X-ray (EDX) spectrometer or analyzer (Horiba Emax energy EX-320) was used to obtain the cross-sectional images of the membrane specimens and carry out their elemental analysis.

3. Results and discussion

3.1. Permeation of pore-fill type palladium membrane

The porous YSZ tube modified with thin YSZ layer (ca. 1.5 μm) was provided as the precursor for the preparation of pore-fill type palladium membranes. Figure 3 shows the cross-sectional SEM micrograph of the pore-fill type membrane along with the EDX line analysis of palladium. Palladium deposited mainly in the pore networks of the thin YSZ layer located under the top surface. Almost no nitrogen leak detected through the obtained membranes indicated that they were sufficiently gastight.

Figure 3. Cross-sectional SEM micrograph of the pore-fill type palladium membrane.

3.2. Permeation experiments

Figure 4 shows hydrogen permeance profiles for the pore-fill type palladium membranes at 300–600°C. The membranes were prepared by different electroless plating solutions including 0.00 wt.%, 0.02 wt.%, and 0.05 wt.% of Triton® X-100. The hydrogen permeance increased with increasing the additive amount of Triton® X-100, and reached 3.42 μmol m^{-2} s^{-1} Pa^{-1} at 600°C when 0.05 wt.% of Triton® X-100 was added (Membrane PF-3). Due to the difficult infiltration of the plating solution to the nano-size pores of the YSZ layer, palladium particles were sometimes deposited on the top surface; which caused the increase in the gas transfer resistance of the membrane. Triton® X-100 could suppress this deposition to the top surface by decrease in the surface tension of the plating solution.

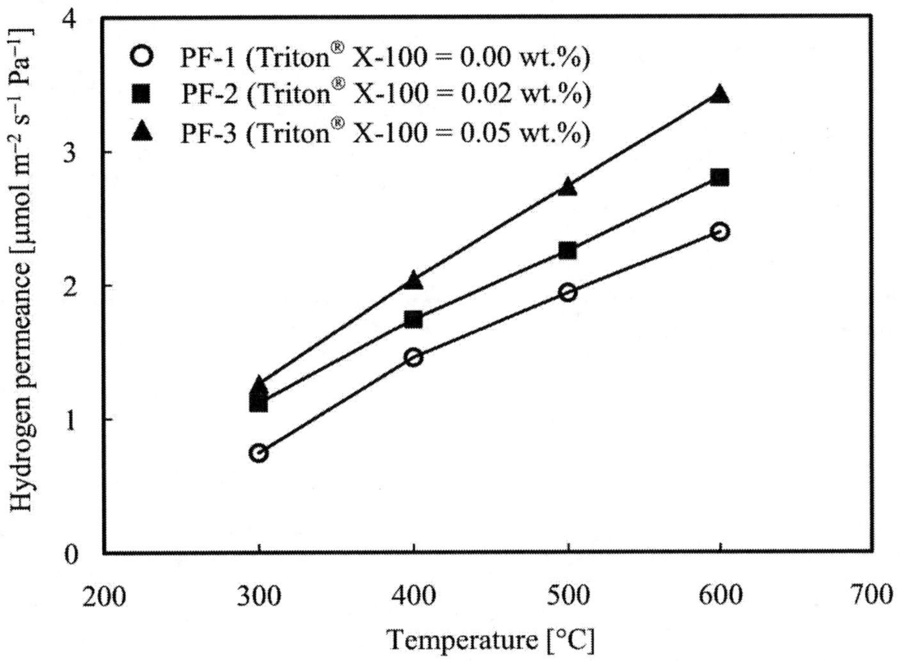

Figure 4. Hydrogen permeance profiles for the pore-fill type palladium membranes at 300–600°C with transmembrane pressure differences of 50 kPa.

Table 1 summarizes the effect of Triton® X-100 on the hydrogen perm-selectivity of the pore-fill type palladium membranes at 600°C. The Hydrogen permeance as well as the hydrogen selectivity of the membranes was enhanced by the addition of Triton® X-100 due to the dramatic decrease in the nitrogen leak. It indicates that nano-size pores of the YSZ tube were effectively plugged by palladium particles with the addition of Triton® X-100.

The long-term gas permeation test was demonstrated with pore-fill type palladium membrane at 600°C under 50 kPa of transmembrane pressure difference; which was prepared under the same condition as the membrane PF-3. In the course of the test, supply gas was alternately switched between hydrogen and nitrogen to confirm the gas tightness of the membrane. The hydrogen permeance of the membrane exceeded 2.11 μmol m^{-2} s^{-1} Pa^{-1} during the 100 h. The hydrogen selectivity was almost constant around 100 due to a small nitrogen leak. However, this nitrogen leak was virtually unchanged during the permeation test. This indicates that the membrane could be available for the high temperature hydrogen separation without a significant fracture of the gas separation layer.

Table 1. Addition effect of Triton® X-100 to the electroless plating solution on the performance of the pore-fill type palladium membranes.

Membrane	Triton® X-100 concentration [wt.%]	H_2 Permeance* [μmol m^{-2} s^{-1} Pa^{-1}]	H_2/N_2 Selectivity*
PF-1	0.00	2.39	1200
PF-2	0.02	2.80	16000
PF-3	0.05	3.42	20000

*The data was obtained by gas permeation tests at 600°C.

4. Conclusion

The fabrication conditions of pore-fill type palladium membrane were optimized to improve the permeance and selectivity for hydrogen. The hydrogen/nitrogen selectivity of the membrane increased by addition of Triton® X-100 to the electroless plating solution, and reached around 20000 at 600°C when 0.05 wt.% of the surfactant was added. The hydrogen permeance reached 3.42 μmol m^{-2} s^{-1} Pa^{-1} at the same experimental conditions. The membrane also showed the good thermal stability during the gas permeation test at 600°C.

Acknowledgements

We would like to thank Dr. Margot A. Llosa Tanco, Mr. Onbhuddha Patiwat, Ms. Mettaya Kitiwan, and Mr. Shuichi Tanno for their fruitful discussions and supports.

References

[1] S. N. Paglieri and J. D. Way, Innovations in palladium membrane research, *Sep. Purif. Methods,* 2002, **31**, 1–169.

[2] S. Uemiya, Brief review of steam reforming using a metal membrane reactor, *Top. Catal.,* 2004, **29**, 79–84.

[3] D. W. Lee, Y. G. Lee, S. E. Nam, S. K. Ihm and K. H. Lee, Study on the variation of morphology and separation behavior of the stainless steel supported membranes at high temperature, *J. Membr. Sci.,* 2003, **220**, 137–153.

[4] D. A. Pacheco Tanaka, M. A. Llosa Tanco, T. Nagase, J. Okazaki, Y. Wakui, F. Mizukami and T. M. Suzuki, Fabrication of hydrogen-permeable composite membranes packed with palladium nanoparticles, *Adv. Mater.,* 2006, **18**, 630–632.

[5] D. A. Pacheco Tanaka, M. A. Llosa Tanco, J. Okazaki, Y. Wakui, F. Mizukami and T. M. Suzuki, Preparation of "pore-fill" type Pd–YSZ–α-Al_2O_3 composite membrane supported on α-Al_2O_3 tube for hydrogen separation, *J. Membr. Sci.,* 2008, **320**, 436–441.

[6] R. J. Wolf, M. W. Lee and J. R. Ray, Pressure-composition isotherms for nanocrystalline palladium hydride, *Phys. Rev. Lett.,* 1994, **73**, 557–560.

[7] K.J. Bryden and J.Y. Ying, Nanostructured palladium membrane synthesis by magnetron sputtering, *Mater. Sci. Eng. A,* 1995, **204**, 140–145.

[8] K.J. Bryden and J.Y. Ying, Nanostructured palladium-iron membranes for hydrogen separation and membrane hydrogenation reactions, *J. Membr. Sci.,* 2002, **203**, 29–42.

[9] N. Q. Minh, Ceramic fuel cells, *J. Am. Ceram. Soc.,* 1993, **76(3)**, 563–588.

[10] J. Kim and Y. S. Lin, Synthesis and oxygen-permeation properties of thin YSZ/Pd composite membranes, *AIChE J.,* 2000, **48(8)**, 1521–1529.

[11] D. J. Green, P. Z. Cai and G. L. Messing, Residual stress in alumina-zirconia laminates, *J. Eur. Ceram. Soc.,* 1999, **19**, 2511–2517.

[12] R. Kojima, K. Sato, T. Nagano and Y. Iwamoto, Development of fine porous alumina capillaries by a dry-wet spinning method, *J. Ceram. Soc. Jpn.,* 2006, **114(11)**, 929–933.

(8e) Development of Hitz Zeolite Membrane for Acetic Acid Dehydration

Kentaro Shinoya, Masanobu Aizawa, Suguru Fujita, Yoshinobu Takaki, Kazuhiro Yano, Takehiro Shimizu, Yoshihiro Asari, Jun Yano, Takaharu Yagi and Ken-ichi Sawamura, Business and Product Development Center, Hitachi Zosen Corporation, Osaka, Japan

Development of Hitz Zeolite Membrane for Acetic Acid Dehydration

Kentaro SHINOYA, Masanobu. AIZAWA, Suguru FUJITA, Yoshinobu TAKAKI, Kazuhiro YANO, Takehiro SHIMIZU, Yoshihiro ASARI, Jun YANO, Takaharu YAGI and Ken-ichi SAWAMURA

Hitachi Zosen Corporation, 2-11, Funamachi 2-Chome, Taisho-ku, Osaka 551-0022, JAPAN

Acetic acid (AcOH) is one of the important chemical products. However its aqueous solution as an intermediate in the production process possesses azeotropic characteristic and the separation of water/ AcOH by distillation requires a large amount of heat energy due to the low relative volatility of them. Membrane dehydration process with high acid resistance would be most candidate to solve the problem and especially mordenite type zeolite (MOR) has been studied for the acid dehydration purpose, because MOR with high Si/Al ratio in the chemical formula is expected to possess high acid resistance. In this background, MOR membrane has been developed in Hitz using a secondary-growth method. As for the substrate, α-Al2O3 porous tube whose size of 16 mm in diameter and 60 mm in length has been adopted. It was found that Hitz MOR membrane exhibited high separation performance, the permeation flux was ca. 2.0 kg m^{-2} h^{-1} for the feed of AcOH (80 wt%) / H2O (20 wt%) mixture and the separation factor exceed 1000 at 75 degree centigrade in PV condition.

See more of this Session: YP Tutorial: Separations
See more of this Group/Topical: Separations Division - Please See Topical 8

(8f) Application of Hitz Zeolite Membranes for Removal of Water and CO_2 From Air

Ken-ichi Sawamura, Yoshihiro Asari, Takehiro Shimizu and Masanobu Aizawa, Business and Product Development Center, Hitachi Zosen Corporation, Osaka, Japan

Membrane separation will offer us great opportunity to deliver more compact and less capital-intensive processes, in comparison with conventional separation technologies such as distillation, extraction, and adsorption. So far, Hitachi Zosen Corporation (Hitz) has successfully developed zeolite membranes with both high flux and high selectivity, resulting in the commercial application for bio-ethanol plant with the capacity of 50,000 L/d in Japan, 2008. The developed membranes, or NaA-type and NaY-type zeolite membranes, showed excellent separation performance in bio-ethanol production and solvents purification; however, performances in other applications are yet to be clarified.

In this study, we applied the zeolite membranes to removal of water and CO_2 from air for production of purified dry air required in various industries. The system we have developed successfully provides dry air with CO_2 concentration of less than 100 ppm. Detailed permeation and separation properties of the zeolite membranes will be presented.

See more of this Session: YP Tutorial: Separations
See more of this Group/Topical: Separations Division - Please See Topical 8

(14a) The Energy Challenge: Providing Sustainable Supplies to Meet Growing Demand

F. Emil Jacobs, ExxonMobil Research and Engineering Company, Annandale, NJ

As economies grow and living standards improve, the world faces a dual energy challenge: meeting the growing demand for energy while mitigating emissions and their impact on the environment. ExxonMobil believes that addressing this dual challenge requires an integrated set of solutions which includes increasing the supply of all commercially viable energy sources, reducing emissions, and improving energy efficiency in all sectors of the economy. Technology will be a key enabler of continued progress in all of these areas.

Many experts agree that the immense scale of global energy demand necessitates a continuing role for oil and gas as a substantial portion of the energy supply mix for at least the next several decades. These experts also forecast that the overall energy mix will shift toward a relatively higher level of sustainability through a combination of increased efficiency across the economy, growth in lower-carbon sources of energy, and development and deployment of technologies that help mitigate emissions.

ExxonMobil has developed a number of technologies that contribute to increased supplies of energy, reduced emissions, and higher efficiency in its operations and in consumer use of its products. The presentation will provide an overview of ongoing research and development activity in a broad range of technology areas including algae-based biofuels. It will also highlight some of company's current technologies including lithium-ion battery separator films, lightweight plastics that improve automotive efficiency, and processes that could advance carbon capture and storage.

See more of this Session: Spring Meeting Keynote Address - Sponsored by the AIChE Fuels & Petrochemicals Division
See more of this Group/Topical: Liaison Functions

(75a) PSA In Biomass to Fischer-Tropsch Fuels Processes with CO_2 Capture

Ana Mafalda Ribeiro, Jo„o Carlos Santos and Alírio Egídio Rodrigues, Laboratory of Separation and Reaction Engineering, Department of Chemical Engineering, Faculty of Engineering, University of Porto, 4200-465 Porto, Portugal

The use of biomass as a source of renewable energy has received much attention lately. The conversion of biomass into valuable energy can be accomplished through a variety of technologies from which gasification appears to be the most promising and efficient one. Gasification converts biomass into synthesis gas that can be further processed into a wide range of high value products such as methanol, hydrogen and Fischer-Tropsch products. However, prior to the synthesis process, the bio-syngas must be conditioned, which includes the removal of carbon dioxide for subsequent sequestration and capture. The most common technologies used for CO_2 removal from syngas are absorbent-based systems such as the Rectisol and Selexol processes. In this study a PSA process is suggested for the replacement of the absorbent-based CO_2 removal process. A pressure swing adsorption cycle was developed for the simultaneous production of a carbon dioxide enriched stream and a (H2+CO) stream with the required specifications of the Fischer-Tropsch reactor feed. An activated carbon was selected as adsorbent. Simulation results showed that it was possible to obtain a CO_2 product with purity above 95% and with 90% recovery. Additionally a H_2/CO stoichiometric ratio of 2.14 was obtained for the produced syngas stream which is adequate for further processing in the Fischer-Tropsch reactor. The power consumption of the CO_2 capture was found to be of 3.36 MW. This value represents a reduction of about 28 % when compared to a Rectisol process with the same CO_2 recovery.

See more of this Session: Gas Separation Technologies
See more of this Group/Topical: Topical 6: 11th Topical Conference on Gas Utilization

(75b) Modified 'Molecular-Gate' PSA for CO_2 Capture From Flue Gas

Raghwendra Singh Thakur[1], Nitin Kaistha[1] and D. P. Rao[2], (1)Department of Chemical Engineering, Indian Institute of Technology Kanpur, Kanpur, India, (2)Process Intensification Consultant, Hyderabad, India

Keller II and Kuo in 1982 proposed a dual-piston Pressure Swing Adsorption (PSA) for separation of mixtures and termed it as *molecular gate* PSA. It consisted of a single adsorbent bed connected with two piston-cylinder assemblies at the ends of the bed to impose cyclic flow of gas with feed entering at an intermediate location along the bed. High and low pressures in the bed are varied by manipulating the displacement volume of pistons. They claim that high purities of both products in the range 99 mol% and high productivity can be achieved at moderate vacuum. However, its application is limited by the size of piston-cylinder assembly that could be used in practice, for example, for CO_2 capture from flue gas. To overcome this limitation, we propose a modified molecular gate PSA replacing the piston-cylinders with two surge tanks, a compressor, a vacuum pump and mass flow controllers. The cycle essentially has only two steps, pressurization and depressurization in which cyclic flow of gas is maintained through the bed. The lead/lag functionality of pistons in molecular gate PSA has been realized using the mass flow controllers. An additional modification of molecular gate has been proposed in timing products withdrawal by passing tanks during the steps to enhance purity and productivity. We have carried out simulation studies on separation of synthetic flue gas containing 20 mol% CO_2 and 80 mol% N_2, using Zeolite 13x. The results indicate that it is possible to get CO_2 purity and recovery in excess of 99% with high pressure of 1 bar and low pressure 0.08 bar with productivity of 43.9 LSTP/h/kg a performance superior to the performance of conventional PSA. The productivity can be further enhanced by 3.5 times upon relaxing the purity specification from 99 mol % to 95 mol%.

Keywords: modified molecular gate, CO_2 capture, flue gas

Reference:

Keller II, G.E. and Kuo, C.H.A.: Enhanced Gas Separation by Selective Adsorption, US Patent 4,354,859 (1982).

See more of this Session: Gas Separation Technologies
See more of this Group/Topical: Topical 6: 11th Topical Conference on Gas Utilization

PSA SEPARATION OF CH4/N2 MIXTURE WITH CARBON MOLECULAR SIEVE

Fang Xi, Wensheng Lin, Anzhong Gu*

*(*Corresponding author: linwsh@sjtu.edu.cn)*

Institute of Refrigeration and Cryogenics, Shanghai Jiao Tong University, Shanghai, China

Abstract

Coalbed methane (CBM) is a kind of unconventional natural gas, and liquefaction is a good alternative to make use of it. In China, CBM often contains air when it comes out of the well. Considering issues regarding safety, it is natural that oxygen must be removed firstly. Then, the effective enrichment of methane from methane-nitrogen mixture is one of the most important issues in the liquefaction of CBM. In this paper, a two-bed pressure swing adsorption (PSA) experimental apparatus was used to separate CH_4/N_2 mixture on a selected type of carbon molecular sieve (CMS) adsorbent. Owing to the kinetic separation mechanism on CMS, nitrogen is adsorbed on CMS, while methane with certain pressure is delivered continuously to the next process of liquefaction. The influence of adsorption pressure, feed gas velocity and cycle time on the overall performance of the adsorption process was discussed. Feeding a mixture of 50% CH_4/50% N_2, methane purity of 95.45% was obtained; and 94.89% purity was obtained while feeding a mixture of 30% CH_4/70% N_2. The results show that all three parameters may affect the separation performance, and the later two have greater influences. It is easier to achieve methane purity above 90% at low adsorption pressure and low feed gas velocity. And the shorter the cycle time, the higher the methane purity is.

Key words: methane-nitrogen mixture; pressure swing adsorption; carbon molecular sieve; coal-bed methane

Introduction

Methane contained in the coal bed methane (CBM) is a kind of clean fuel, but its emission to atmosphere would contribute to the greenhouse effect. The methane content in CBM is usually only 20% to 40% due to the mixing of air during the exploitation process[1]. In spite of the large coal production, CBM is not well utilized in China. Besides its low methane content, CBM wells are often remote and the gas production of a single well is so small that delivering it through pipeline is economically unpractical. An effective way of utilizing CBM is to deliver it to local natural gas supply systems after liquefaction[2].

For oxygen mixed in CBM, it must be removed firstly considering the safety issue. So the major and difficult task of utilization CBM is to separate methane from nitrogen economically. Although cryogenic distillation is a mature application, it usually needs higher energy consumption for the undesired liquefaction of nitrogen. An alternative way of separating the CH_4/N_2 mixture is pressure swing adsorption (PSA) process by using the mechanism of kinetic selectivity to produce enriched methane with certain pressure.

Principally, a kinetic separation is achieved by virtue of the difference in diffusion rates which is higher for the smaller nitrogen molecular compared to methane. Although simple in principle, it is a relatively difficult separation compared to an equilibrium separation because of the low values of the diffusion time constants and the adverse equilibrium selectivity towards methane and nitrogen. Ackley [3] enriched methane from 50% to 80% by vacuum pressure swing adsorption (VPSA). However, it is hard to enrich methane up to 90% through kinetic PSA while the methane content is lower than 50% in the feed gas until now[4]. The selection of appropriate adsorbent is crucial to the successful separation of methane and nitrogen.

In our recent study[5], single bed kinetic PSA separation experiments were carried out under a wide range of adsorption pressures(P), feed gas velocities(v) and different methane contents in feed gas with a selected carbon molecular sieve (CMS). The breakthrough curves showed that at the initial stage of adsorption there was only methane at the outlet of adsorption bed, indicating that methane could be enriched up to 90% with a single PSA unit.

In this paper, the same two simulated methane-nitrogen mixtures (30% and 50% methane, respectively) were separated using a two-bed PSA unit with the same CMS adsorbent. The experimental studies under various conditions of adsorption pressures, feed gas velocities and cycle times were investigated.

Experimental

A simple two-bed PSA process operated in a cyclic manner as shown in *Figure 1* comprises four basic steps: repressurization, feed, counter-current blowdown and counter-current vacuumization. High pressure feed flows through bed B where the faster diffusing nitrogen is preferentially adsorbed, leaving a stream richer in methane at the outlet of the bed after the regenerated bed repressurized with product. Vacuum regeneration is used to remove most of the nitrogen which has been adsorbed in order to avoid slow uptake of methane which would saturate the adsorbent and reduce its effectiveness for taking up nitrogen during the subsequent adsorption step.

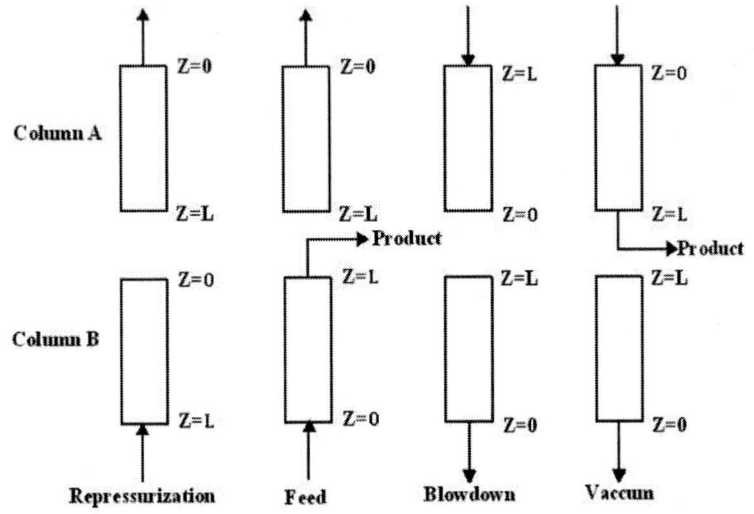

Figure 1 Schematic diagram of steps involved in PSA cycle

Table 1 Experimental conditions of the two-bed PSA unit

Feed gas composition	50%CH_4, 50%N_2
	or 30%CH_4, 70%N_2
Column length	30cm
Column diameter	3cm
Adsorbent	CMS
Mass of adsorbent in each column	136g
Particle size	φ1.5×3 cylinder
High pressure	0.3 or 0.6MPa(Abs.)
Velocity of feed gas	0.2 or 0.4 cm/s

The experimental unit is shown in *Figure 2*. The two stainless steel columns were packed with a known amout of CMS(Table 1), which was produced by Changxing Shanli Chemical Material Technology Co.,Ltd. Activation of CMS was done at 423K overnight under high vacuum conditon. Each column was connected to three pneumatic valves: one for the inlet of feed gas, one for the outlet of product in feed step and the inlet of product gas in repressurization step, and the other for the outlet of absorbed nitrogen in blowdown and vacuumization steps. The pneumatic valve, which was used to prestect the experimental apparatus from possible explosion, was driven by pressurized air. The circulation of the pressurized air was controlled by two-way solenoid valve. Sequential switching of solenoid valves was controlled by relays. An electronic mass flow controller (MFC)(±0.3%) was used to controll the feed gas flow. The product and feed stream are analysed by withdrawing gas samples through a manual sampling probe and submitting them to a GC900A chromatograph. The experimental unit was evacuated for 12 hours and purged with pure helium before each experiment. The pressure of the unit was controled by a back pressure valve.

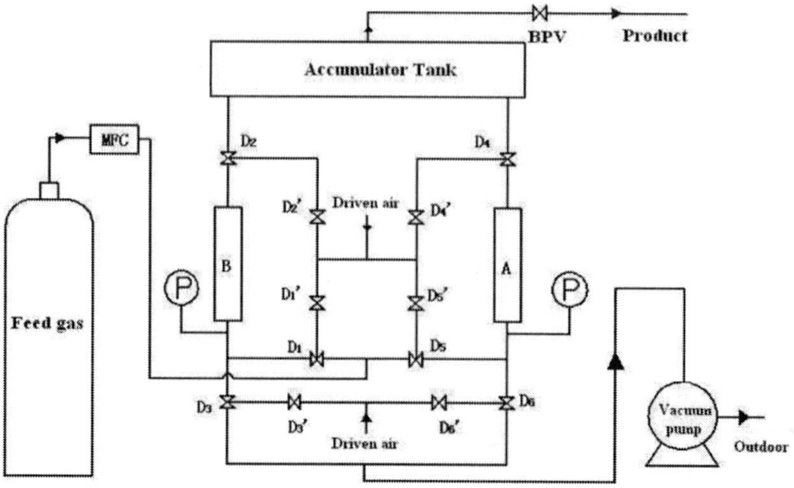

D_1, D_2, D_3, D_4, D_5, D_6: solenoid valve; D_1', D_2', D_3', D_4', D_5', D_6': pneumatic valve;

MFC: mass flow controller; A, B: column; BPV: back pressure valve; P: pressure gauge.

Figure 2 Schematic diagram of the two-bed PSA experimental unit

Results and discussion

It has been proved that velocity of feed gas, adsorption pressure, and cycle time can affect the performance of a PSA unit[6]. Our recent study[5] showed that the greater the velocity of feed gas was, the lower the methane purity was in product while the velocity was between 0.2-1.0cm/s. In this two-bed PSA research, velocities of 0.2 and 0.4cm/s were chosen according to the single-bed experiments. The adsorption pressures studied were 0.2 and 0.5MPa.

The unit was purged and pressurized by helium at the beginning of experiments, so there are two kinds of curves in the methane concentration plots as showed in *Figure 3*. The solid ones were the methane concentration in methane and nitrogen binary mixture(*CH4%2*) in buffer tank, manually eliminated helium. This profile represented the separation effect of the CMS directly. The hollow ones were the methane concentration in methane, nitrogen and helium ternary mixture($CH_4\%3$), which were the true methane concentration of the product in buffer tank, representing the productivity of adsorbent. In these two figures, *t* was the adsorption time of a single bed. All the concentration points correspond to those measured 2s prior to the end of each cycle.

The results show that the methane concentration is enriched to a certain extent under each condition. While *v* is 0.2cm/s, $CH_4\%2$ is up to 90% under all experimental conditions, and the highest one is 95.45%. But it is different while *v* is 0.4cm/s. $CH_4\%2$ is up to 90% only when the methane concentration in feed gas($CH_4\%1$) is 50%, and the highest one is 94.87%. Table 2 is the summary of experiments.

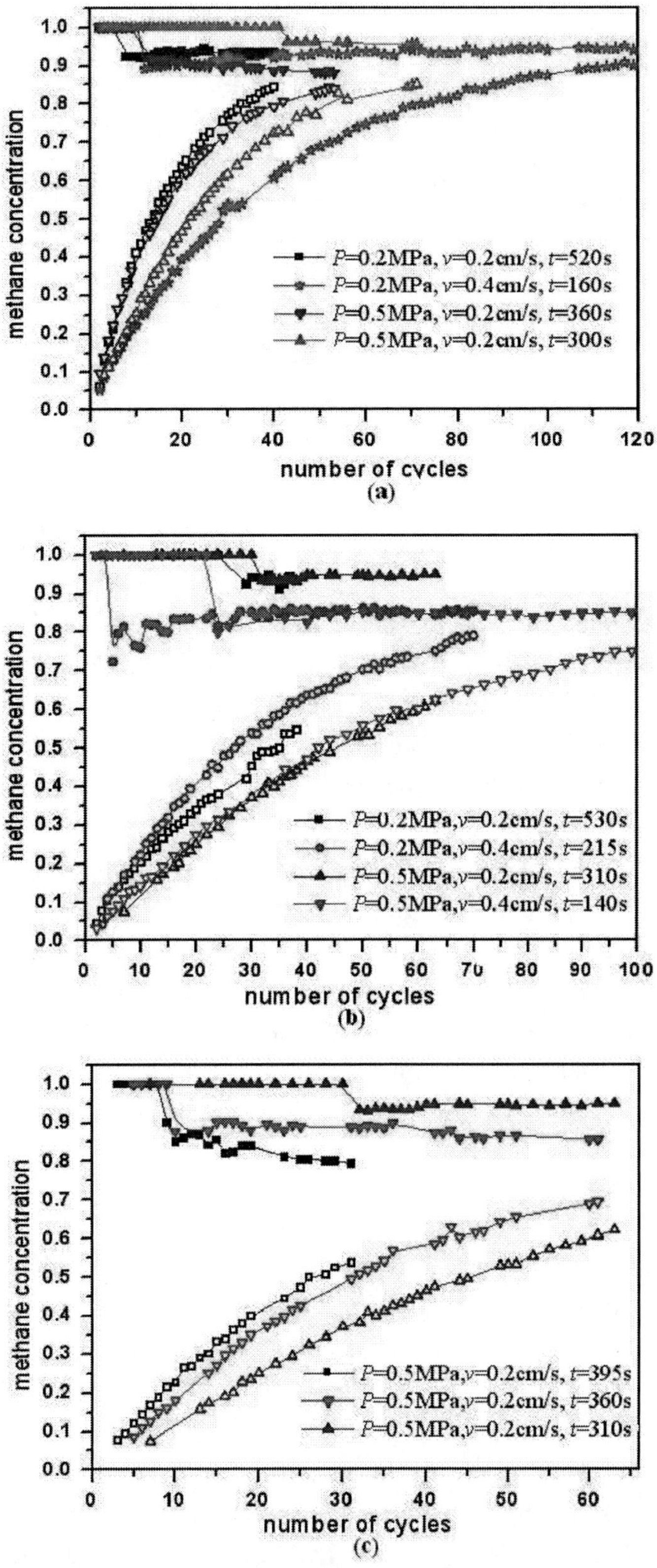

Figure 3 Methane concentration profiles while the methane concentration was 50%(a) and 30%(b, c) in feed gas balanced with nitrogen.

Table 2 Summary of two-bed PSA experiments with CMS (temperature: 300K)

P (MPa)	*v* (cm/s)	$CH_4\%1$ (V%)	*t* (s)	$CH_4\%2$ (V%)	$CH_4\%3$ (V%)
0.2	0.2	50	520	93.64	84.54
		30	530	93.23	54.76
	0.4	50	160	94.37	90.78
		30	215	85.67	79.05
0.5	0.2	50	360	88.04	84.21
			300	95.45	85.01
		30	395	79.53	53.62
			360	85.96	69.44
			310	94.89	62.18
	[illegible]	[illegible]	[illegible]	[illegible]	[illegible]

It has to be noted that the reason of the slow increase of $CH_4\%3$ is the much greater volume of buffer tank compared with the volume of column, in order to make sure that the pressure decline of the system after repressurization is small enough. It is interesting that $CH_4\%2$ is 100% at the beginning of the adsorption. In fact, the nitrogen extent is too small to be analyzed though GC900A. In the enlarged profiles from the chromatograph, there is a chromatographic peak of nitrogen.

It can be seen from the results that cycle time has a great influence on methane purity. Methane purity in the product is higher while the cycle time is shorter no matter what other experiment conditions are. The influences of adsorption pressure and feed gas velocity are the same as that in the single-bed study: the increases of both parameters have negative effects on the methane purity. As can be seen from the profiles, while the methane concentration in feed gas is 30%, the accumulation of methane in buffer tank is slower than that when methane concentration is 50%.

Conclusions

Experimental PSA studies were performed with two methane-nitrogen feeds (50% and 30% methane in nitrogen) using a two-bed unit packed with a selected CMS. A four-step PSA cycle was used. The exit methane concentration ranged from 95.45% to 88.04% for the 50% feed and from 94.89% to 79.53% for the 30% feed.

The cycle time, feed gas velocity and adsorption pressure could affect the methane purity, and the methane concentration in feed gas significantly influences the accumulation speed of methane in buffer tank. It is easier to achieve methane purity above 90% at low pressure and low velocity. And cycle time has a negative effect on methane purity.

The study shows that it is possible to enrich methane concentration from a relatively low level to a commercially acceptable one, implying potential application in CBM utilization practices.

References

[1] Nie LH , Xu SP , Su YM , Liu SQ. Progress of recovery of low concentration coal bed methane. Chemical Industry and Engineering Progress, 2008,27(10):1505-1511.

[2] Lin WS, GU M, GU AZ, et al. Analysis of coal bed methane enrichment and liquefaction processes in China. In: Proceedings of the 15th International Conference & Exhibition on Liquefied Natural Gas. Barcelona, Spain, 2007.

[3] Ackley M W, Yang R T. Kinetic separation by pressure swing adsorption: Method of characteristics model. AIChE Journal, 1990, 36 (8):1229-1238

[4] Guo P, Li M. Research progress of separation of CH_4/N_2 in coal-bed methane. Chemical Industry and Engineering Progress, 2008, 27(7):963-967.

[5] Xi F, Lin WS, Gu AZ. Separation of methane/nitrogen mixture by pressure swing adsorption on carbon molecular sieve. Journal of the China Coal Society, in press.

[6] Fatehi A. Separation of methane nitrogen mixtures by pressure swing adsorption using a carbon molecular sieve. Gas Separation & Purification, 1995,31(9):199-204.

(75d) Two-Dimensional Modeling the Transfer Phenomena In the Radial Adsorber

Xingang Zheng, **Yingshu Liu**, Wenhai Liu, Yongling Li and Hui Zhang, School of mechanical engineering, University of science and technology Beijing, Beijing, China

In this study, a two-dimensional model is used to study the transfer phenomena occurred in the radial adsorber with consideration of heat effect, gas compressibility and dead volumes. The distributions of concentration, temperature and velocity in four possible configurations, CP-z, Cp-Π, CF-z, and CF-Π of a radial flow adsorber are investigated. As expected, the results visually show the evolution of process variables and their distribution. Some discussions of the effect of different configurations on variables distribution are made. The two-dimensional simulation is found to be very useful for a better understanding of the dynamic sorption and separation process in the radial adsorber.

See more of this Session: Gas Separation Technologies
See more of this Group/Topical: Topical 6: 11th Topical Conference on Gas Utilization

(86a) Modeling Flux Decay In Cross-Flow Filtration

Philip P. Schonewill, Richard C. Daniel, Rick W. Shimskey and Reid A. Peterson, Pacific Northwest National Laboratory, Richland, WA

Recent work at Pacific Northwest National Laboratory has been concerned with understanding how cross-flow filter flux decays with time. Understanding the cause(s) of flux decay is important for assessing long-term behavior of this separation technology, especially when used in challenging processing environments e.g. nuclear waste treatment. Fouling in the filter is a complex issue which is affected by many parameters: the chemistry of the solution being filtered, particle size and morphology, operating conditions, and cleaning methods. Based on filtration work performed at PNNL over the past several years, it is clear that a portion of the fouling can be "reversed" by back-pulsing the filter and another portion is irreversible (usually only rectified via cleaning protocols). However, the sheer number of parameters requiring consideration complicates effective modeling of both the long-time behavior of the cross-flow filter and its response to back-pulses. To combat this, a series of simple models were developed from first principles based on different suspected fouling mechanisms. The models were tested against long-term data collected with PNNL's bench-scale filtration system. These models suggest some simple relationships between operating parameters and flux decay, provide a framework for predicting performance for other, similar filtration systems, and assist in determining a back-pulse strategy for cross-flow filters which will experience long-term decays in the filter flux.

See more of this Session: Advances In Fluid Particles Separation
See more of this Group/Topical: Separations Division - Please See Topical 8

(86b) Particulate Separation In Drilling Process and Filter Cake Formation: CFD Modeling and Simulations

Isaac K. Gamwo, National Energy Technology Laboratory, U.S. Department of Energy, Pittsburgh, PA and Mohd A. Kabir, U.S Department of Energy, National Energy Technology Laboratory, Pittsburgh, PA

Oil and gas producers are drilling wells to ever increasing depths to maintain reserves and production. Downhole pressures and temperatures increase with well depth, as a result, the properties of the drilling fluid vary. One of the most important functions of these fluids is to seal off the face of the wellbore so that the fluid is not loss to the formation. Ideally, this is accomplished by depositing a filter cake of the particulate in the drilling fluid over the surface of the wellbore without a loss of solids to the formation. This is critical in conducting initial drilling into production formation.

In this study, an Eulerian-Eulerian multiphase model is used to simulate the multiphase flow and particulate separation process at the wellbore wall as filter cake is formed. The effect of particle sizes on the filter cake thickness has been investigated. Filter cake formation for both shallow and ultra deep depth drilling conditions have been modeled and simulated.

See more of this Session: Advances In Fluid Particles Separation
See more of this Group/Topical: Separations Division - Please See Topical 8

DE-OXYGENATION EXPERIENCE AT WORLD'S LARGEST SEAWATER TREATMENT PLANT FOR OIL RESERVOIR INJECTION

Dr. Nicos P. Isaias, Mohammed H. Al-Ghamdi, Ibrahim B. Din, Ahmad M. Hawsawi
Sea Water Injection Department, Saudi Aramco, Dhahran, Saudi Arabia

Abstract

Seawater used for oil reservoir injection should be disinfected and should also be non-corrosive, non-scaling, and free of particulate matter to minimize oil reservoir damage. A Saudi Aramco seawater treatment plant facility, (Plant #1), with a present design capacity of 14 million barrels per day, is the world's largest seawater treatment plant for oil reservoir injection. The plant disinfects the seawater, for bacterial control, through the addition of sodium hypochlorite upstream of the deaeration process and by batch biocide addition, downstream. The suspended solids are removed through granular media filtration using the world's largest horizontal deep bed pressure filters. The dissolved oxygen of the filtered seawater is removed by cascading it down the de-oxygenation tower, where the seawater splashes onto a series of trays, causing dissolved oxygen to be lost to a nitrogen stream, which is blown counter-currently. Sulfurous acid is added to scavenge the remaining oxygen to control corrosion and prevent aerobic bacterial growth. Nitrogen and sulfurous acid are both manufactured on-site. Residual sulfurous acid is maintained in the seawater transport pipelines to avoid seawater re-oxygenation.

The objective of this paper is to give an overview of the treatment plant process and, in particular, to describe in some detail, the de-oxygenation process, the improvements made to the system and the challenges encountered.

Keywords

Seawater, Nitrogen Stripping, Sulfurous Acid, De-oxygenation, Reservoir Injection

Introduction

Plant #1, a Saudi Aramco facility in the Arabian Gulf, is the world's largest seawater treatment plant for oil reservoir injection. It has a present design capacity of 14 million barrels per day (MMBD) of treated seawater of which, 12.5 MMBD is used for injection into producing reservoirs to enhance oil recovery.

To minimize formation damage to the reservoir, the seawater is disinfected for bacterial control, is filtered to remove suspended solids to avoid plugging the tiny pores of the reservoir rock, and is deaerated to remove dissolved oxygen to avoid corrosion to downstream equipment and to control aerobic bacterial growth.

The injection of water into the reservoir rock is necessary to increase pressure and thereby to increase oil production. This method is known as waterflooding and increases oil production in two ways:

1. Water injected at the outer edges of the reservoir sweeps or displaces oil from the reservoir towards the producing wells.
2. Water injection helps in maintaining a constant reservoir pressure, thereby maintaining the oil production rate over a longer period of time.

Seawater is preferred for injection water rather than using aquifer water or produced water. Aquifer water in Saudi Arabia is water that has been geologically entrapped and is therefore of a finite quantity as there is not enough rainfall in Saudi Arabia to replenish it. Although aquifer water has the advantage of purity, it needs to be preserved for future generations rather than be used in large quantities for oil reservoir injection. Although produced water reduces the potential of formation damage from the compatibility point of view; it enhances the risk of scaling and corrosion in the injection pipelines. Produced water also requires treatment to remove hydrocarbons and solids to render it suitable for re-injection. Furthermore, the volumes of water being produced are never sufficient to replace all the production volumes and additional mixing with "make-up" water may be required, thereby further enhancing the risk of scaling. Seawater is relatively clean and is of uniform composition. Placing of the seawater intake at a sufficient depth eliminates algae. The high seawater salinity (57,000 mg/l at Plant #1) also helps to avoid clay swelling.

The main disadvantages of seawater are that it requires disinfection for bacterial control, de-oxygenation to reduce corrosion related problems, and perhaps its greatest drawback is its scaling tendency[1]. Therefore, the filtered water must meet stringent water quality requirements. Table 1 shows the seawater characteristics and the final treated seawater quality specifications targeted at Saudi Aramco's Plant #1 treatment facility.

Table 1. Seawater characteristics and treated seawater specifications

Seawater Characteristics	Symbol	Untreated Seawater	Treated Seawater
Total dissolved solids, (mg/l)	TDS	57,000	57,000
Total suspended solids (mg/l)	TSS	2 - 3	< 0.2
pH, (standard units)	pH	8.1	7.2 - 7.3
Dissolved oxygen, (mg/l	O_2	6.0 - 8.0	< 10 ppb
Sulfite residual, (mg/l)	SO_3	0.0	6.0 - 12.0
Particle size (microns)	CC (Coulter Counts)	4,000 - 7,000 > 2.0 μ	Less than 10% of the particles shall be larger than 2.0 micron size.

Suspended solids at Plant #1 are removed by granular media filtration. Growth of micro-organisms is controlled by injection of sodium hypochlorite, produced on-site by hypochlorite generators and the periodic use of biocides. Oxygen is removed by nitrogen stripping and the residual oxygen is scavenged using sulfurous acid. Both nitrogen and sulfurous acid are manufactured on-site. Removal of oxygen minimizes contamination of the water with corrosion products. The sulfurous acid injection causes the seawater pH to decrease, thereby reducing seawater alkalinity, which prevents calcium carbonate scaling. However, a side effect of the sulfurous acid addition is the removal of the chlorine residual and therefore, periodically biocides need to be injected into the system.

The objective of this paper is to give an overview of this unique seawater treatment facility and in particular to describe, in some detail, the de-oxygenation process and to outline the challenges encountered, as well as the improvements made to the system.

Description of Saudi Aramco Plant #1

Plant #1 receives seawater with a high salinity of approximately 57,000 mg/l of total dissolved solids (TDS) from the Arabian Gulf. Figure 1 shows that the first step is for the seawater to be disinfected by injecting sodium hypochlorite, which is generated on-site using the electrolytic cell process.

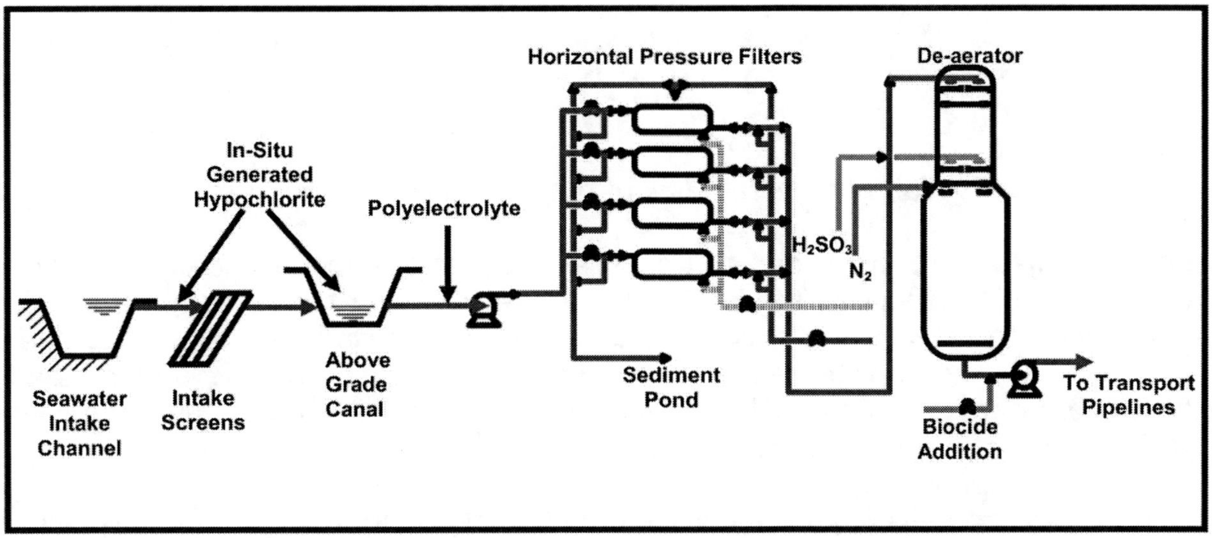

Figure 1. Saudi Aramco Plant #1 process flow diagram.

Sodium hypochlorite is added to the seawater at the intake structure to discourage fish from reaching the plant intake and for bacterial control. Sodium hypochlorite is also added in the above-grade canals to maintain chlorine residual and also to control algae growth on the canal walls. All the marine organisms (large fish, plankton and algae) are removed from the seawater by the use of mechanical screens, such as bar and drum screens. After passing the screens, the seawater is pumped to two above-grade canals. Each canal acts as a reservoir for feeding the horizontal pressure filters. The filter modules and the deaerators are located on either site of the central canals from which they are fed, Figure 2. Each canal feeds 14 treatment modules. Each module has a design capacity of 0.5 million barrels per day and consists of four filters and a deaerator column for oxygen removal. Filtration aid (polyelectrolyte) is added to agglomerate the seawater colloidal particles to enhance the filtration performance.

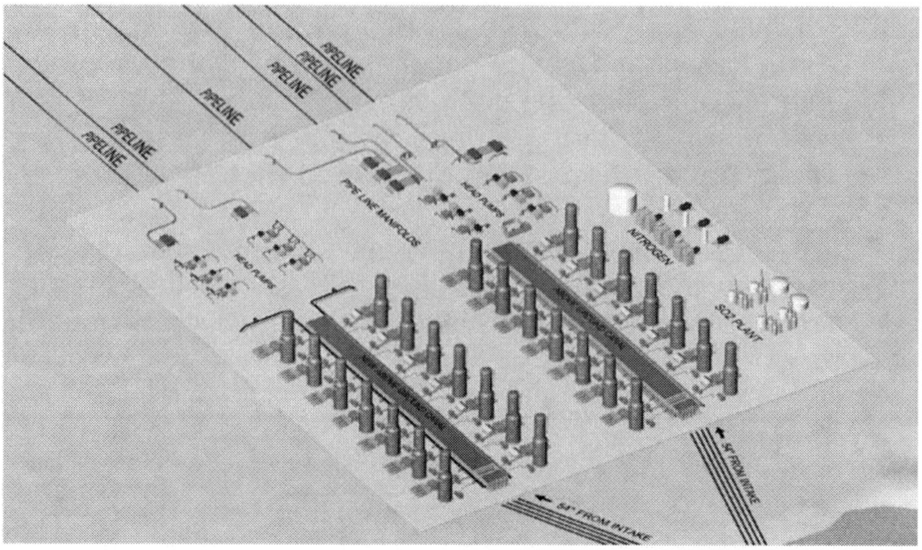

Figure 2. An overall view of Plant #1 seawater treatment.

This chemical treatment is absolutely necessary to achieve the efficiency required of the removal of the high suspended solids. The filters are designed for the removal of 95% of all suspended solids greater than 2 microns and less than 0.2 mg/l in the filters' outlet. To meet this stringent seawater filtrate requirement, the plant has to operate under steady-state conditions. Large and sudden changes in filtration flow rate, which can cause an increase in turbidity in the filtrate water, must be avoided.

The filtrate from the horizontal filters is fed to deaerators where counter-flow nitrogen stripping takes place for the removal of dissolved oxygen. Oxygen needs to be removed from the seawater to control corrosion. The oxygen content in seawater is approximately 6 mg/l and 8 mg/l for summer and winter, respectively. After oxygen removal in the deaerator by the counter-flow of nitrogen, the residual oxygen level is of the order of 0.4 mg/l. Sulfurous acid is then added to scavenge this residual oxygen and to achieve oxygen concentrations as low as <10 ppb, however, a side effect of the sulfurous acid addition is the removal of the residual chlorine. Therefore, the deaerated water has no residual biocidal capability. It is therefore necessary to inject biocide for biological control into the treated seawater after deaeration. The injection of biocide is generally done in one to three slugs per week rather than being added continuously. The treated water is pumped through pipelines as far as 300 km away for reservoir injection.

Deaeration Process Description[2]

De-oxygenation takes place in the deaeration tower. Figure 3 shows typical deaeration towers used for the removal of oxygen from the filtered seawater. The size of the deaeration towers is 85 feet tall, with a 20-foot diameter storage section at the bottom and a 16-foot diameter stripping section on top. Removal of oxygen in the deaeration tower is done in two different steps.

Figure 3. Deaeration towers used for the removal of oxygen from filtered seawater.

a. De-oxygenation by Mechanical Removal

The filtered seawater is fed through the inlet header at the top of the deaeration tower and cascades down the tower through trays. The oxygen content in seawater, which is approximately 6 mg/l and 8 mg/l for summer and winter, respectively, is stripped out from the seawater by nitrogen, blown counter-currently as shown in Figure 4. The oxygen and nitrogen escape to the atmosphere through the vent on top of the tower. This mechanical stripping by nitrogen gas removes 95% of the oxygen from the filtered seawater.

The factors that affect the removal of oxygen from the filtered seawater are:

1. Nitrogen flow rate. A higher nitrogen flow rate will remove more oxygen from water.
2. Flow rate of seawater in the deaerator. Lower seawater flow rate results in higher oxygen removal.
3. Seawater temperature. Higher seawater temperature improves the oxygen removal efficiency because the solubility of oxygen in seawater decreases with increasing temperature. The seawater temperatures experienced are 90 °F - 95 °F and 60 °F - 65 °F for summer and winter, respectively.
4. Purity of the stripping gas nitrogen. As the oxygen increases in the nitrogen gas, the nitrogen has less attraction for the oxygen in the seawater. The purity of nitrogen from the nitrogen plants on-site is high (< 190 ppm of oxygen in nitrogen).

The remaining 5% of oxygen after mechanical deaeration, which corresponds to approximately 0.4 mg/l, is scavenged by chemical means by the addition of a sulfur dioxide solution, which is produced by the sulfur plants.

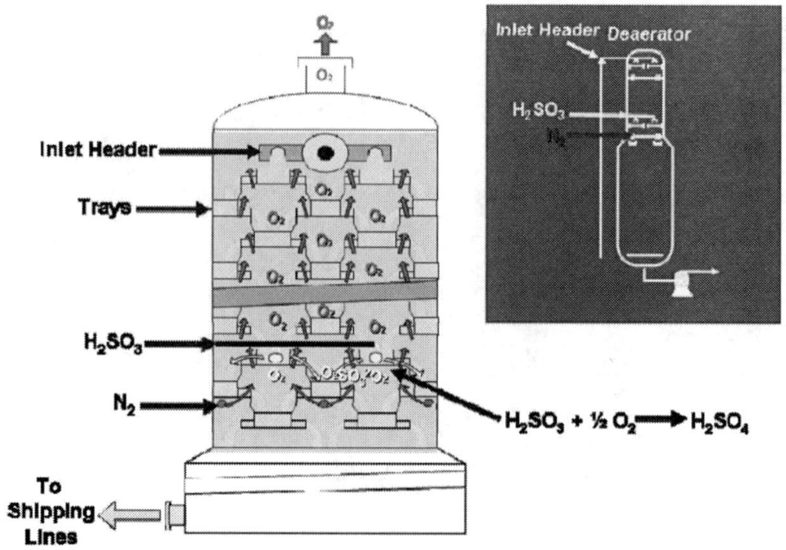

Figure 4. Deaeration tower.

b. De-oxygenation by Chemical Means (Scavenging)

Sulfur dioxide solution manufactured on-site by the sulfur plants is injected into the seawater coming out of the stripper section of the tower to react with the residual oxygen left in the seawater. This oxygen removal is called scavenging and results in no more than 10 ppb oxygen leaving the deaerator tower. A large reservoir at the bottom of the stripper allows the required time (3 minutes) for the chemical reaction between oxygen and sulfur dioxide solution to take place.

The addition of sulfur dioxide solution has a two-fold purpose.

a) To scavenge the residual oxygen and

b) To reduce the seawater pH.

Reduction of the pH value reduces the seawater's scaling potential by alkalinity reduction. However, the sulfur dioxide solution also reacts with chlorine and eliminates all the chlorine residual from the seawater. Therefore, once the seawater leaves the deaerator, there is no residual biocidal capability in the treated seawater. It is therefore necessary to inject a biocide into the treated seawater. The treated seawater is then pumped through pipelines as far as 300 km away for reservoir injection.

Deaeration System Challenges and Improvements

Overall Stripping Efficiency

One of the main challenges of the system is to maintain a high stripping efficiency (as per design at 95% removal of dissolved oxygen). The dissolved oxygen inlet to the deaerator tower is approximately 6 ppm and 8 ppm for summer and winter, respectively. Furthermore, the deaerated seawater flow rate also varies between 400 MBD - 500 MBD per deaerator, depending on the downstream export pipelines requirement. Therefore, frequent field test runs are done rigorously at various seawater loading rates to determine the overall efficiency. The design of the deaerator internal trays also helps in achieving high de-oxygenation efficiency. Sulzer Chemtech A.G.[3] of Switzerland is the proprietary tray design manufacturer. The design employed an angled tray arrangement with four path flows, with 14 trays, a tray spacing of 27 inches, a weir height of 2 inches, and a down-comer clearance of 5 inches. Apart from ensuring a high de-oxygenation efficiency of 95%, this design results in a very low tray pressure drop. Figure 5 shows a typical graph of dissolved oxygen removal vs. the number of trays used in the deaerator design.

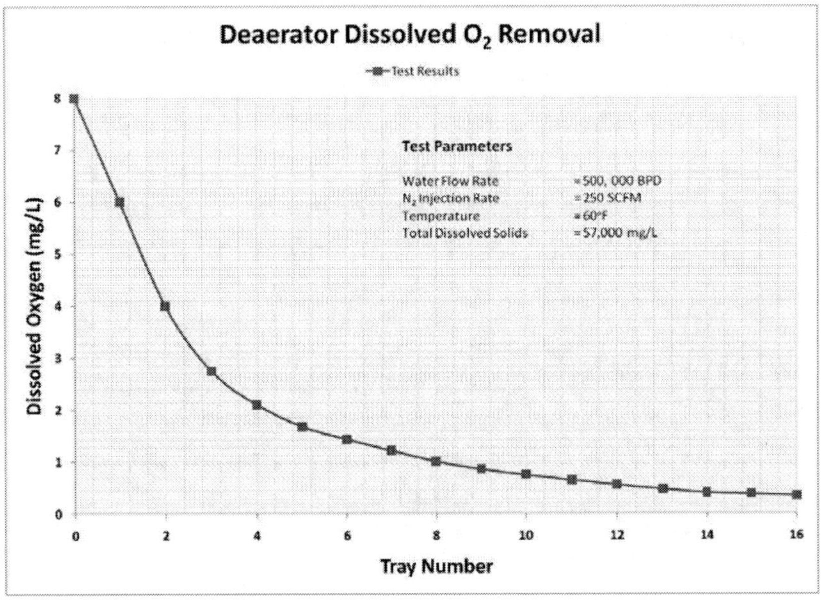

Figure 5. Dissolved oxygen removal vs. number of trays.

Optimization of N_2 Injection Rate

The optimization of nitrogen gas injection is possible after conducting the rigorous field test runs at different vapor loading rates. The optimum nitrogen gas injection rate is between 200 scfm - 250 scfm, compared with 400 scfm as called by the original design[4] with a 95% stripping efficiency. During the test run, close monitoring of dissolved oxygen is necessary at the inlet and outlet of the deaerator liquid streams. The injection of nitrogen stripping gas is also automated fully with flow control valves in place.

Deaerator Pilot Testing

One of the deaerators is equipped to allow pilot testing. This deaerator, which is of the same size as the other deaerators, is very special as it allows sampling from each individual tray. Therefore, each tray's efficiency can be studied. Sight glasses on each tray are also available during the trial to visually inspect the foaming phenomena and stripping process.

pH Control

The pH of normal seawater is around 8.1. The desirable treated seawater pH is 7.2 - 7.3 to avoid scaling in the pipelines and the downstream injection facilities. Optimum pH control is achieved by online pH analyzer monitoring, which automatically adjusts the sulfurous acid injection rate accordingly. The injection of sulfurous acid (oxygen scavenger) is automated fully by a flow control valve.

Optimization of Oxygen Scavenger

The optimization of the oxygen scavenger injection rate is achieved by conducting rigorous field test runs at different seawater flow rates with a fixed nitrogen gas stripping rate. The optimum oxygen scavenger injection rate of between 15 gpm and 20 gpm compared with the design rate of 30 gpm, still yields the 8 mg/l - 12 mg/l sulfite residuals. During the test run, close monitoring of sulfite residual and pH is done at the outlet of the deaerator liquid streams. Figure 6 shows a typical residual oxygen graph vs. reaction time with a sulfur dioxide solution. The graph shows that the dissolved oxygen in the de-oxygenated seawater meets the specification of <10 ppb after 3 minutes reaction time with 12 ppm SO_2 residual.

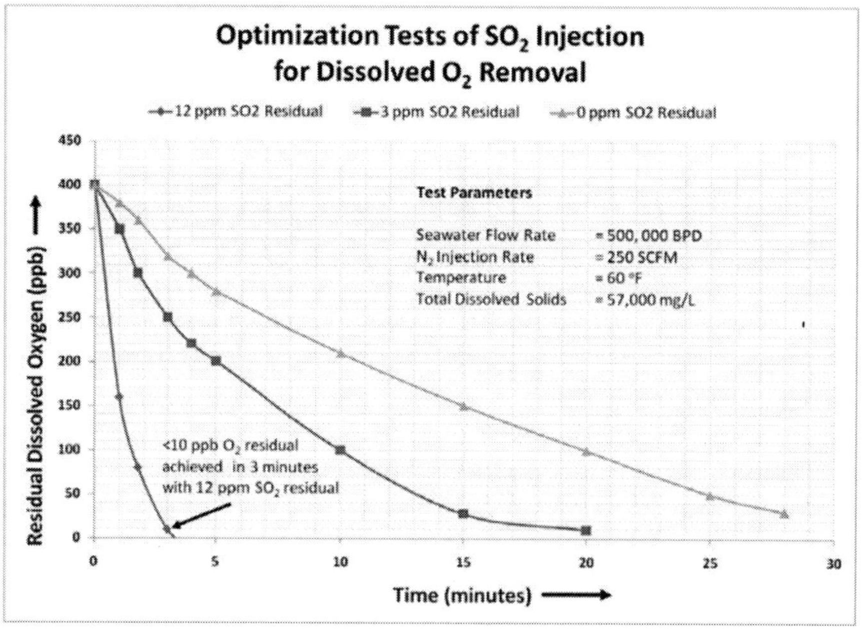

Figure 6. Residual oxygen vs. reaction time with SO_2 solution.

Corrosion Control

Seawater is very corrosive to carbon steel and other lower grade stainless steels. For cost-effective considerations, the carbon steel is used for the body material of the deaerator, which is 16 feet in diameter and 37 feet high for the tray's tripping section. The bottom oxygen scavenging section, which is 20 feet in diameter and 27 feet high, is also made of normal carbon steel material. Therefore, internally the deaerator is epoxy coated at 12 mils dry film thickness for corrosion protection. Galvanic aluminum anodes (sacrificial anodes) are also put in place at strategic points inside the deaerator to ensure corrosion protection.

Nitrogen Gas Entrainment into the Liquid Product

The presence of gas in water is closely monitored by measuring the content of gas vaporized in a specific volume of liquid accumulated. The gas water ratio (GWR) is maintained at 2% - 3% at the inlet of the pipelines. Downstream of the pipelines, at high elevation points, the gas will normally come out of solution due to a hydraulic gradient/pipeline pressure drop and has to be vented out by the vent valves. Optimum and desirable GWR results are obtained through optimizing the vapor loading rates. The nitrogen gas injection rates are regulated optimally to put the tray operating region away from the flooding region. The bottom collection pan design has also been modified to include 7,815 sieves/holes of one-half inch diameter in its four active panels. This can offer a "gentle rain" of seawater of the bottom tray that will prevent severe entrainment. Nitrogen entrainment in the transported pipelines can cause problems with the injection pumps which can trip because of the low suction.

Conclusion

Plant #1 is unique in terms of size and process. Treating large volumes of seawater is very challenging and meeting the seawater quality specifications for oil reservoir injection is even more challenging. Many of the challenges in the de-oxygenation of seawater by the deaerators are successfully addressed. As a result, maximum operational performance and high de-oxygenation efficiency are maintained.

References

1. Mohammed H. Al-Ghamdi, Nicos P. Isaias, "Experience from the World's Largest Seawater Filtration Plant for Oil Reservoir Injection," 10th World Filtration Congress, Leipzig, Germany, April 14-18, 2008.
2. Qurayyah Sea Water Plant (QWSP) Operating Manual, Chapter 12, Section E.
3. Process Data Sheet for Deaerator Trays Qurayyah Sea Water Plant, Project 10-03451-001, Doc. No. BE-863418, pp. 1-5, August 1, 2006.
4. Process Flow Diagram, Deaerator and Bottom Pump Qurayyah Sea Water Plant, Project 10-03451-001, Doc. No. BA-534100, sheet No. 1, Revision No. 2, November 1, 2009.

(86e) Different Methods of Extraction of Natural Colorant From Marigold (Tagetes Erecta L.), Solubility, Purification, and Recuperation of Solvents In Process

Felix Gamarra, Department of Gama, University of Brasilia, UnB Gama College, Brasilia, Brazil, **Elias Tambourgi**, Chemical Engineering, State University of Campinas, Campinas, S„o Paulo, Brazil and **Edison Bittencourt**, Chemical Engineering, State University of Campinas, Campinas, S„o Paulo, Brazil, Brazil

Increasing legislative limitations and restrictions imposed for the world-wide organisms to the use of synthetic colorants, resulted renewed interest for colorants natural, that come back to have importance, and appear as possible substitutes in the industrial of foods, textile, others, diminishing the ambient impact and damages to the human organism. The marigolds, even so the flower extract is used as supplement in the feeding veterinary medicine, traditional form, is scarce and dull the available information in literature in terms of incomes, qualities, stability, security and applications. In this work he is demonstrated that to the method and extraction with supercritical fluid (SFE), using as solvent the supercritical CO_2 is most efficient in the xanthophylls pigment extraction (96%). Also was used more two methods of extraction as the immersion and lixiviation with some changes. The used percentages of recovery of the solvents during the process had all been highly satisfactory, superior 80% in all the cases. The objective of this work is to determine the performance of extracts, using solvent (ethanol, among others), for methods of leaching and SFE, as well as the recovery of the solvents, and purification of lutein.

See more of this Session: Advances In Fluid Particles Separation
See more of this Group/Topical: Separations Division - Please See Topical 8

(120a) Beer Tower Revamp Experience

Daniel R. Summers, Sulzer Chemtech USA, Inc., Tulsa, OK

Revamp experience in a beer distillation tower.

See more of this Session: Bio Fuels Distillation: Looking to the Future
See more of this Group/Topical: Topical 8: The Dr. James Fair Heritage Distillation Symposium

(120b) Alternative Separation Methods for Bioethanol Purification

Aristoklis Hadjitheodorou, School of Chemical and Metallurgical Engineering, University of the Witwatersrand, Johannesburg, South Africa and **Mark Peters**, Wits University, Johannesburg, South Africa

The environmental impacts of fossil fuel usage along with the depletion of available reserves have led scientists and engineers to seek greener and more renewable sources of energy. Bioethanol has proven to be one of the most promising alternatives to fossil fuel-derived petroleum as a transport fuel. Typical ethanol concentrations of fermentation from microorganisms are around 15%. This product is a very dilute solution therefore a large portion of the energy expenditure in the production of bioethanol is used to extract ethanol from the bioreactor effluent. Currently distillation is the most commonly used method of separation, however, it is ineffective in this situation as a large amount of energy is utilised in vapourising water which exits the column mainly as a bottoms product. Distillation is also limited as an azeotrope forms at 95%wt ethanol. Thus it is of significant interest to investigate other common separation methods in bioethanol purification, either as an alternative to distillation or as an addition, in an effort to reduce overall energy consumption. Column Profile Maps (CPM), originally developed for distillation systems, offers a graphical method for rapidly synthesizing complex separation configurations. The CPM technique, which is directly applicable to stripping and absorption processes, has been extended to other separation processes, such as membrane permeation and reactive distillation and can be developed for others (i.e. solvent extraction). Thus feasible hybrid systems can easily be synthesized from a similar basis and compared to the traditional method of distillation in terms of their 'ease of separation' and overall work efficiency.

See more of this Session: Bio Fuels Distillation: Looking to the Future
See more of this Group/Topical: Topical 8: The Dr. James Fair Heritage Distillation Symposium

Ethanol Distribution Over Fractions of Ethanol-Gasoline Fuel From Distillation and Spectroscopic Measurements

Sergey A. Karpov and Natalia B. Teterevyatnikova, Gubkin Russian State University of Oil and Gas, Moscow, Russia

The search for alternative sustainable fuels has grown enormously, especially in the 1990s as attention has been focused on environmental issues. At the forefront currently for additives to motor gasolines is ethanol. Ethanol is renewable and has the potential of reducing carbon dioxide emissions as compared to traditional fuels. It can be used in non-adapted, current vehicles at levels up to 10% by volume.

As it was mentioned above, environmental concerns have lead to the desire to add oxygenates to traditional gasolines. In the US and other parts of the world, methyl tert-butyl ether (MTBE) was accepted as the most economical oxygenate. Unfortunately, MTBE entered groundwater in some areas of the US, and its very low threshold for taste and odor have rendered the water undesirable for human consumption. The removal of MTBE opens a gap for an environmentally acceptable high-octane component for gasoline and is a key driving force in the search for alternative components. Ethanol is emerging as the additive to fill the MTBE gap and, it is being phased into California and the rest of USA where oxygenated fuels are required.

The addition of ethanol to hydrocarbons can significantly affect the volatility properties. Mixtures of ethanol with hydrocarbons exhibit large deviations from ideal mixing as defined by Raoult's law. Sometimes such non-ideal behavior can be reasonably predicted for binary systems using standard activity coefficient models and newly developed equations of state applicable for associating molecules (e.g. CPA and SAFT).

With the exception of octane number specifications, the most important properties defining current "boutique" gasolines are volatility properties. The key volatility properties are Reid vapor pressure and distillation.

It follows from the above-stated that it is necessary to understand the structure of alcohol containing gasoline components in order to resolve the problem of their excessive volatility.

It is obvious that when adding alcohol to gasoline it forms a number of azeotropes with gasoline hydrocarbons. Such azeotropes determine excessive volatility of alcohol containing gasoline components in comparison with the source gasoline. In this connection, it should be considered that different azeotropes between hydrocarbons were present in the system before adding of alcohol to the system.

Because of a complex fractional composition of gasoline we may not expect exact knowledge of composition and structure of all azeotropes, being formed by ethanol. At the same time, we may expect some general regularity. Particularly, a link should probably exist between group composition of the source gasoline and increase in vapor pressure when adding alcohol.

In order to find these general regularities, it is necessary to determine, how ethanol added is distributed between different fractions of gasoline.

The Pasquini paper describes the use of near infrared (NIR) spectroscopy for real-time monitoring of distillations. The high scan speed of an AOTF-based NIR spectrometer has been combined with a simple flow cell designed to collect the condensate whose absorbance spectrum is remotely acquired in real time during the distillation, through a pair of optical fiber bundles. The potential of the proposed system has been demonstrated via its use in the distillation of some laboratory-prepared binary mixtures of hydrocarbons, hydrocarbon and ethanol, and regular and adulterated gasoline samples.

Unfortunately NIR signal cannot be used for quantitative analysis of fuel fractions.

In Ref.[7] it is shown that, among the oxygenated compounds used as additives for gasolines, ethanol was the one that caused the most marked change in the distillation curve, through the formation of a threshold due to the minimum temperature azeotrope that was formed. The nonoxygenated compounds, isooctane and toluene, hardly altered the distillation curve.

Lack of quantitative model of ethanol distribution impedes all subsequent discussions at this stage. So this work is related to ethanol distribution over fractions of alcohol containing fuel.

Distillation of alcohol containing gasoline was performed in compliance with ASTM D 86 using manual distillation apparatus HDA 620 (HERZOR, Austria). This type of devices cannot be called the best for component separation, but their use is a standard in petroleum and related industries. Their results are used for estimation of quality of gasoline and other "light" petroleum products. So we have decided to use this type of device instead of any other (even more effective).

Distillation was performed at 5 °C intervals. Since – as our experience and Ref.[7] showed – ethanol significantly changes the shape of distillation curve, increasing its "sharpness" within 60-70 °C, the temperature range was varied from time to time with a view to collect a sufficient amount of fractions for subsequent analysis.

Determination of ethanol content in hydrocarbon mixture is possible using both gas chromatography method8 and infrared spectroscopy method. A good correlation between these two methods was demonstrated.0

As IR spectroscopy method is not so time consuming and doesn't require use of additional reagents, its use for analysis of 112 fractions, obtained in this work, is preferred. Moreover, application of attenuated total reflectance (ATR) add-on device makes the analysis even easier.

IR spectroscopy method was used for determination of ethanol content in gasoline and its fractions. Analysis was performed using Tensor 27 FT-IR Spectrometer (Bruker, Germany). MIRacle ATR add-on device (Bruker, Germany) with ZnSe crystal was used for analysis of fuel fractions. Detailed information on the unit and add-on device is given in Table 1 [3]. The obtained fractions were stored at the temperature of -18 °C not more than 24 hours since the moment of distillation.

Figure 2 [3] gives examples of distillation curve, obtained by us. As it is evident from this figure, relationships are analogous to those obtained before (see Ref.[1] and Fig. in Ref.[7]).

Addition of ethanol significantly changes distillation curves of gasoline. Ethanol leads to essential change in the curve shape – a plateau appears at about 68 °C. Furthermore, ethyl alcohol leads to essential increase in the quantity of low-boiling-point fractions. For example, if in the source gasoline (Gasoline A), whose distillation curve is given in Figure 2 [3] under 70 degrees Celsius 25% v/v of volume boiled off; after addition of 5% v/v ethanol under 70 degrees Celsius 51% v/v began to boil off. By increase in alcohol percentage to 10% v/v, this percentage increased to 56% v/v.

It was already found1 that the D-86 is similar to the TBP, having a "plateau" region during the ethanol–hydrocarbon azeotropes "region". The "step change" is now a "transition region" due to the sloppy separation of the D-86. The D-86 for our gasoline–ethanol blends has three distinct regions: plateau or azeotropic region, transition region, and dilution only region as already observed by authors of Ref.[1].

Figure 3 [3] illustrates influence of alcohol on distillation temperatures at percentage evaporated. One can see that addition of ethanol mostly influence on T10 and T50 points. The same situation is observed for Gasoline B. One should say that the change in alcohol concentration results in significant change of mostly T10 and T50.

Main objective of our work was to determine how alcohol is distributed over gasoline fractions. From Figure 2 [3] it is evident that the main part of alcohol isn't a clean gasoline component (characterized by own boiling point11 – 78. °C at 101. kPa). If alcohol didn't form azeotropes with gasoline hydrocarbons, its addition would result in a simple increase in quantity of fraction, boiling off within 75-80 °C. But the main influence of ethanol addition is below than that temperature limit.

This is not observed in Figure 2 [3]. Figure 2 [3] shows that alcohol forms a number of azeotropes with gasoline components, where most of these azeotropes has obviously a lower boiling point than the source hydrocarbons.

Examples of IR spectra (500-4000 cm-1) of two different fractions (with mean boiling temperatures of 47. and 77. °C) of alcohol containing fuel (Gasoline B + 10% v/v ethanol) are given in Figure 4 [3]. Figure 4 [3] shows that the maximum signal within the range of 3200-3600 cm-1 is caused by «light» gasoline fractions. Fractions with average boiling point over 90 °C have only traces of alcohol. These amounts are under our detection error for ethanol (0.% v/v), therefore we accepted the alcohol content of these fractions as zero.

The same fact was observed in Ref.[6] (see Fig. 6) by NIR method but no quantitative analysis was provided there.

Figure 5 [3] gives distribution of alcohol over different fractions of alcohol–gasoline composition. The maximal amount of ethanol is found in fraction 65-70 °C – up to 30% v/v. Low ethanol concentration is in the lowest boiling part of gasoline (IB-42 °C) and 85-90 °C – less then 5% v/v.

Attention is drawn to the fact of low alcohol content in its «native» fraction (with = 77. °C). So we can conclude that the main volume of alcohol is found in azeotrope form. It determinates the high volatility of ethanol–gasoline mixtures.

We showed in this work how exactly ethanol is distributed over different fractions of alcohol–gasoline blend. To our knowledge, this result is the first of its kind. Only qualitative data have been published previously (see "Introduction").

We found out that ethanol is concentrated in light fractions and is almost completely absent in heavy fractions (with boiling point over 90 °C). Its traces, found in fractions with high boiling points, should be probably referred to indistinct component separation during distillation as per ASTM D 86 (without reflux). The distribution form is like Gaussian distribution with maximum at ~60 °C.

Main part of ethanol is found to form azeotropes – the reason of the fuel high volatility. Only small part of alcohol (~3% v/v) "stays" in its "natural" form.

In the following, the use of these data14 – as we hope – will contribute to more precise determination of nature of increased vapor pressure for alcohol–gasoline blends and to solution of this problem, which will make alcohol–gasoline blends more competitive as compared to traditional gasolines.

References

1. S. A. Karpov (2008), "Aliphatic alcohols as components of fuels for internal combustion engines"

Chemistry and Technology of Fuels and Oils 44, 249.
2. S. A. Karpov (2008), "Current aspects of fuel ethanol production in Russia and abroad" Chemistry and Technology of Fuels and Oils 44, 1.
3. R. M. Balabin, R. Z. Syunyaev, S. A. Karpov (2007), "Molar enthalpy of vaporization of ethanol-gasoline mixtures and their colloid state" Fuel 86, 323.
4. R. M. Balabin, R. Z. Safieva (2007), "Capabilities of near infrared spectroscopy for the determination of petroleum macromolecule content in aromatic solutions" J. Near Infrared Spec. 15, 343.
5. Z. Nan, Z.-C. Tan, and L. Sun (2004), "Investigation on Thermodynamic Properties of Ethanol + Gasoline Blended Fuel" Energy Fuels 18, 84.
6. N. Peschke, S. I. Sandler (1995) "Liquid-Liquid Equilibria of Fuel Oxygenate + Water + Hydrocarbon Mixtures. 1" J. Chem. Eng. Data 40, 315.
7. S. Hellinger, S. I. Sandler (1995) "Liquid-Liquid Equilibria of Fuel Oxygenate + Water + Hydrocarbon Mixtures. 2" J. Chem. Eng. Data 40, 321.
8. R. M. Balabin (2010), "Is quantum chemical treatment of biopolymers accurate? Intramolecular basis set superposition error" J. Chem. Phys. 132, 231101.
9. R. M. Balabin, R. Z. Safieva, E. I. Lomakina (2010), "Gasoline classification using near infrared (NIR) spectroscopy data: Comparison of multivariate techniques" Anal. Chim. Acta 671, 27.
10. R. M. Balabin (2010), "Intramolecular basis set superposition error as a measure of basis set incompleteness: Can one reach the basis set limit without extrapolation?" J. Chem. Phys. 132, 211103.
11. R. M. Balabin (2010), "The identification of the two missing conformers of gas-phase alanine: a jet-cooled Raman spectroscopy study" Phys. Chem. Chem. Phys. 12, 5980.
12. R. M. Balabin (2010), "Boryl substitution of acetaldehyde makes it an enol: Inconsistency between Gn/CBS and ab initio/DFT Data" J. Phys. Chem. A 114, 3698.
13. R. M. Balabin (2010), "Conformational equilibrium in glycine: Experimental jet-cooled Raman spectrum" J. Phys. Chem. Lett. 1, 20.
14. M. B. Gramajo de Doz, C. M. Bonatti and H. N. Sólimo (2003), "Liquid–liquid equilibria of ternary and quaternary systems with two hydrocarbons, an alcohol, and water at 303. K: Systems containing 2,2,4-trimethylpentane, toluene, methanol, and water, or 2,2,4-trimethylpentane, toluene, ethanol, and water" Fluid Phase Equilibria 205, 53.
15. M. B. Gramajo de Doz, C. M. Bonatti and H. N. Sólimo (2003) "(Liquid+liquid) equilibria of ternary and quaternary systems with two hydrocarbons, an alcohol, and water at T=303. K. Systems containing cyclohexane, benzene, methanol, and water" The Journal of Chemical Thermodynamics 35, 825.
16. Jasem A. Alkandary, Adel S. Aljimaz, Mohamed S. Fandary and Mohamed A. Fahim (2001) "Liquid–liquid equilibria of water + MTBE + reformate " Fluid Phase Equilibria 187-188, 131.
17. R. M. Balabin (2009), "Tautomeric equilibrium and hydrogen shifts in tetrazole and triazoles: Focal-point analysis and ab initio limit" J. Chem. Phys. 131, 154307.
18. R. M. Balabin (2009), "Conformational equilibrium in glycine: Focal-point analysis and ab initio limit" Chem. Phys. Lett. 479, 195.
19. R. Z. Syunyaev, R. M. Balabin, I. S. Akhatov, J. O. Safieva (2009), "Adsorption of petroleum asphaltenes onto reservoir rock sands studied by near-infrared (NIR) spectroscopy" Energy Fuels 23, 1230.
20. R. M. Balabin, E. I. Lomakina (2009), "Neural network approach to quantum-chemistry data: Accurate prediction of density functional theory energies" J. Chem. Phys. 131, 074104.
21. R. M. Balabin (2009), "Polar (acyclic) isomer of formic acid dimer: Gas-phase Raman spectroscopy study and thermodynamic parameters" J. Phys. Chem. A 113, 4910.
22. R. M. Balabin (2009), "Enthalpy difference between conformations of normal alkanes: Raman spectroscopy study of n-pentane and n-butane" J. Phys. Chem. A 113, 1012.
23. J. Chen, L.-P. Duan, J.-G. Mi, W.-Y. Fei and Z.-C. Li (2000) "Liquid–liquid equilibria of multi-component systems including n-hexane, n-octane, benzene, toluene, xylene and sulfolane at 298. K and atmospheric pressure" Fluid Phase Equilibria 173, 109.
24. F. Aiouache and S. Goto "Liquid–liquid equilibria of ternary 2M1B/2M2B–2M1BOH–H2O and quaternary 2M1B–2M2B–2M1BOH–H2O mixtures" Fluid Phase Equilibria187-188, 415.
25. R. M. Balabin (2008), "Enthalpy difference between conformations of normal alkanes: Intramolecular basis set superposition error (BSSE) in the case of n-butane and n-hexane" J. Chem. Phys. 129, 164101.

26. R. M. Balabin (2008), "Intermolecular dispersion interactions of normal alkanes with rare gas atoms: van der Waals complexes of n-pentane with helium, neon, and argon" Chem. Phys. 352, 267.
27. R. M. Balabin, R. Z. Safieva, E. I. Lomakina (2008), "Wavelet neural network (WNN) approach for calibration model building based on gasoline near infrared (NIR) spectra" Chemometr. Intell. Lab. Syst. 93, 58.
28. R. M. Balabin, R. Z. Safieva (2008), "Motor oil classification by base stock and viscosity based on near infrared (NIR) spectroscopy data" Fuel 87, 2745.
29. R. M. Balabin, R. Z. Syunyaev (2008), "Petroleum resins adsorption onto quartz sand: Near infrared (NIR) spectroscopy study" J. Colloid Interf. Sci. 318, 167.
30. R. M. Balabin, R. Z. Safieva (2008), "Gasoline classification by source and type based on near infrared (NIR) spectroscopy data" Fuel 87, 1096.
31. I. Hatzioannidis, E. C. Voutsas, E. Lois, and D. P. Tassios (1998) "Measurement and Prediction of Reid Vapor Pressure of Gasoline in the Presence of Additives" J. Chem. Eng. Data 43, 386.
32. R. A. Harley and S. C. Coulter-Burke (2000), "Relating Liquid Fuel and Headspace Vapor Composition for California Reformulated Gasoline Samples Containing Ethanol" Environ. Sci. Technol. 34, 4088.
33. R. M. Balabin, "The First Step in Glycine Solvation: The Glycine-Water Complex" J. Phys. Chem. B 114, 15075 (2010).
34. R. M. Balabin, R. Z. Syunyaev, T. Schmid, J. Stadler, E. I. Lomakina, R. Zenobi (2011), "Asphaltene adsorption onto an iron surface: Combined near infrared (NIR), Raman, and AFM study of the kinetics, thermodynamics, and layer structure" Energy Fuels 25, 189.
35. R. M. Balabin, E. I. Lomakina, R. Z. Safieva, (2011) "Neural network (ANN) approach to biodiesel analysis: Analysis of biodiesel density, kinematic viscosity, methanol and water contents using near infrared (NIR) spectroscopy" Fuel, in press; doi: 10.1016/j.fuel.2010.11.038.
36. E. S. Mortikov, L. Kh. Kunashev, S.A. Karpov, M.A. Brykin (2009), "Preparation of high-octane components of gasoline from methanol over zeolite catalysts" Theoretical Foundations of Chemical Engineering 43, 779.
37. R. M. Balabin, R. Z. Syunyaev, S. A. Karpov (2007), "Quantitative measurement of ethanol distribution over fractions of ethanol-gasoline fuel" Energy Fuels 21, 2460.
38. R. M. Balabin, R. Z. Safieva, E. I. Lomakina (2007), "Comparison of linear and nonlinear calibration models based on near infrared (NIR) spectroscopy data for gasoline properties prediction" Chemometr. Intell. Lab. Syst. 88, 183.
39. A. V. Tsarev, S.A. Karpov (2009), "Increase of the environmental and operational characteristics of automobile gasolines with the introduction of oxygenates" Theoretical Foundations of Chemical Engineering 43, 563.

(120d) Reducing the Energy Demand of Cellulose Based Fuel Ethanol Through Salt Extractive Distillation Enabled by Electrodialysis

Mohammed A. Hussain and Peter H. Pfromm, Chemical Engineering, Kansas State University, Manhattan, KS

The Renewable Fuels Standard (RFS2), established under the Energy Independence and Security Act (EISA) of 2007, mandates the production of 36 billion gallons/year of renewable fuels in the U.S. in 2022, out of which 16 billion gallons/year must be produced from cellulosic sources. The fermentation based corn-to-fuel ethanol process is widely used to produce fuel ethanol in the U.S. While fermentation can be used to produce fuel ethanol from cellulosic sources, the concentration of ethanol in the fermentation broth may be less than 5 wt% compared to 10-15 wt% in case of corn-ethanol. Similar to corn-ethanol production, distillation can be used for separating and concentrating ethanol from fermentation broth followed by molecular sieve adsorption to achieve fuel grade purity, but the separation energy demand increases due to the drastic increase in distillation energy demand for ethanol concentrations less than 5wt%. 25,100 Btu/gal for 5 wt% ethanol compared to 15,100 Btu/gal for 15 wt % ethanol has been reported.[1] Dilute ethanol from the fermentation broth can be separated and concentrated aided by salt extractive distillation to directly produce fuel ethanol leading to significant energy savings. Techniques other than evaporation and drying for recovering salt, which is used to facilitate distillation, have rarely been considered. Since evaporative salt concentration/crystallization and solids drying techniques are energy intensive, reducing the energy demand for the salt recovery step becomes essential for reducing the overall energy demand. In this study, a combination of electrodialysis and spray drying is investigated for recovering salt. The salt extractive column bottoms stream is pre-concentrated by electrodialysis followed by spray drying. In electrodialysis, the dilute salt solution is concentrated by selectively removing the salt ions from the solution rather than evaporating water; therefore, requiring less energy than an evaporative process. Final recovery of dry salt is achieved in a spray dryer. Salt extractive distillation, with salt recovery enabled by a new scheme of electrodialysis and spray drying, is here conceptually integrated in the water-ethanol separation train of a fermentation based cellulosic fuel ethanol plant and investigated through process simulation using Aspen Plus® V7.2. Concepts, modeling, and results showing potential energy savings will be discussed.

[1]- Côté P, Noël G, Moore S. The Chatham demonstration: From design to operation of a 20 m^3/d membrane-based ethanol dewatering system. *Desalination.* 2010;250:1060-1066.

See more of this Session: Bio Fuels Distillation: Looking to the Future
See more of this Group/Topical: Topical 8: The Dr. James Fair Heritage Distillation Symposium

SIMULATION OF ETBE SYNTHESIS FROM TERT-BUTANOL AND ETHANOL BY BATCH RREACTIVE DISTILLATION USING HOMOGENEOUS CATALYSTS

Nada Bahjat Nakkash, Nadeen Khalid Al-Janabi
Al-Nahrain University, Baghdad – Iraq

Abstract

The present work concerned with studying the behavior of batch reactive packed distillation column to produce Ethyl tert-butyl ether (ETBE) by the reaction of tert-butanol (TBA) and Ethanol using two homogeneous catalysts H_2SO_4 and $KHSO_4$, experimentally and theoretically.

Experimentally, after checking the performance of catalysts and their ability to produce ETBE by bench experiment the reaction was carried out in a batch reactive packed distillation unit consists of heat resistance glass column of 71 cm total height and 3.5 cm inside diameter packed with glass rasching ring of 3 mm inside diameter at one atmosphere pressure.

The effect of many operating conditions on ETBE purity such as feed molar ratio of ethanol to tert-butanol (1:1, 2:1, 4:1), reflux ratios (3, 4, and 5), heat duties of (65, 90, and 146W), type of catalyst H_2SO_4 and $KHSO_4$, and the amount of catalyst have been studied, when the catalyst and reactants are mixed in the still. The best operating conditions when the reaction takes place in the still are FMR 1:1, reflux ratio 5, heat duty 146W and H_2SO_4 catalyst with 61.317% purity of ETBE and 94.789% conversion which are higher than using $KHSO_4$ catalyst even when three times its amount was used. For the best operating conditions an experiment was carried out using an intermediate reactive section with $KHSO_4$ catalyst placed in the middle section of distillation column. The results give a purity of ETBE 79.407% and 99.465% conversion which are higher than that obtained from reactive still but it needs large amounts of catalyst.

Theoretically, an equilibrium model (EQ) was developed using MATLAB (R2009b) to solve **MESHR** equations. The developed model results were compared with the results of an existed program ASPEN PLUS (10.2) for certain operating conditions, the deviation between the two programs is about ± 10% error. Results of experimental and theoretical model have been compared and show good agreement.

Keywords: ETBE synthesis, Simulation, Batch Reactive Distillation, Packed distillation.

Introduction

Reactive distillation (RD) is an alternative to a separated reaction-distillation processes, whereby reaction and separation take place with in a single counter current column. Reactants are converted to products in a reaction zone with simultaneous separation of the products and recycle of unused reactants to the reaction zone [1]. Among suitable reactive distillation processes is etherification process to produce octane enhancer such as production of ethyl tert-butyl ether (ETBE), methyl tert-butyl ether (MTBE), tert-amyl methyl ether (TAME), tert-amyl ethyl ether (TAEE) which are the best known oxygenates to be used to increase the octane level and to promote cleaner burning of gasoline and thus decrease harmful emissions from vehicles [2]. There is increasing interest in the use of ETBE as a replacement of methyl tert-butyl ether (MTBE) because of the latter's environmental problems [3], also ETBE has lower blend Reid vapor pressure (4 psi) than MTBE (9 psi) which allows ETBE to be used successfully in obtaining gasoline with less blend Reid vapor pressure than (7.8 psi) as required in some hot places during summer [4]. Furthermore, from environmental view point, ETBE is derived from ethanol which can be obtained from renewable resources like biomass [4].

Mostly ETBE has been synthesized by exothermic reversible reaction between isobutene (IB) and ethanol (ETOH), but the availability of isobutene is limited, it is only produced in refinery using catalytic and steam cracking operations, and it produces ETBE under high pressure. Therefore, alternative routes to synthesize ETBE are under substantial consideration. The most important substitute of isobutene is tert-butanol (TBA) which is a by-product of propylene oxide production in ARCO process.

In the synthesis of ETBE from TBA and ETOH, following sequence of reactions take place [5].

$$\underset{TBA}{(CH_3)_3COH} + \underset{ETOH}{CH_3CH_2OH} \longleftrightarrow \underset{ETBE}{(CH_3)_3COCH_2CH_3} + \underset{Water}{H_2O} \quad ...(1)$$

The above main reaction is accompanied by two side reactions

$$\underset{TBA}{(CH_3)_3COH} \longleftrightarrow \underset{IB}{(CH_3)_2C{=}CH_2} + \underset{Water}{H_2O} \quad ...(2)$$

$$\underset{IB}{(CH_3)_2C{=}CH_2} + \underset{ETOH}{CH_3CH_2OH} \longleftrightarrow \underset{ETBE}{(CH_3)_3COCH_2CH_3} \quad ...(3)$$

As isobutene exists only in the gaseous state at atmospheric pressure, so backward reaction in reaction (2) and reaction (3) can safely be neglected [4, 6].

Batch reactive distillation is suitable when the reaction products have lower boiling point temperature than the reactants [7]. For the present system (ETOH-TBA-ETBE-H_2O), ETBE has the lower boiling point therefore; in the present work batch reactive distillation process is considered.

Matouq et al. [8] synthesized ethyl tert-butyl ether from tert-butanol and ethanol in the presence of different acid catalysts ($KHSO_4$, $NaHSO_4$, H_2SO_4, and Amberlyst 15) at low alcohol grade (mixture of 80% mole percent water).

Sneesby et al. [9] presented a dynamic simulation model for ETBE synthesis from isobutene and ethanol. The simulation was developed using Speed Up simulation software, then the system was utilized for the study of transient open-loop responces and control system design.

Quitain et al. [10] synthesized ETBE from tert-butanol. Then, the effect of feed molar ratio of ETOH/TBA and type of catalyst on conversion and selectivity were studied. The experimental results were compared with results from ASPEN

PLUS simulator and it was shown good agreement.

Bisowarno and Tade [11] used dynamic simulation model to understand the effect of start up policy on the ETBE reactive distillation. This start up policy would results in the targeted ETBE purity within the shortest possible time without using excessive energy.

Yang et al. [12] developed a dynamic mathematical model for synthesis of ETBE from tert-butanol using ASPEN PLUS. The model incorporated reaction kinetics. The rapid solution for this model was obtained using Newton-Raphson method. Their simulation results showed good agreement with the literature results.

Jhon and Lee [13] presented dynamic simulation model using MATLAB software for reactive distillation column of 20 stages with ETBE synthesis. A structured and simple iterative algorithm was devised to estimate the vapor flows with the fast convergence under the rigorous energy balance. Both total and partial condensers are considered, and the reactive distillation performance for these condensers was observed. By comparison of their influence on purity of ETBE, total condenser is the better one.

Assabumrungrat et al. [4] used the experimental results of packed reactive distillation column for synthesis of ETBE from TBA to validate a simulation model using ASPEN PLUS software. They investigated the effect of various operating parameters such as condenser temperature; feed molar flow rate, reflux ratio, heat duty, and molar ratio of H_2O: ETOH on the reactive distillation performance.

Varisli and Dogu [14] produced tert-amyl ethyl ether (TAEE) and tert-amyl alcohol (TAA) by simultaneous etherification and hydration of 2-methyl-2-butene in a batch reactive distillation column. The effect of changing reboiler temperature in the range of (90-124) C° on total conversion was studied.

Hamza [15] studied a batch reactive distillation in packed column for esertification of methanol with acetic acid to produce methyl acetate using sulfuric acid as a homogeneous catalyst. The influence of various parameters such as batch time, reflux ratio, and feed molar ratio of methanol to acetic acid on performance of batch reactive distillation was studied, and the best conditions found are reflux ratio 2, feed molar ratio 2, and batch time 90mit.

Umar et al. [16] used a packed reactive distillation column to generate experimental data of ETBE synthesis from TBA. The effect of different key variables on reactant conversion and product purity in distillate was investigated. Umar et al. [16] studied the experimental synthesis of the same system with different macro porous and gelular ion exchange resin catalyst. Effect of various parameters such as temperature, reactants feed molar ratio, and catalyst loading were studied for optimization of reaction condition.

Nakkash et al. [17] presented a batch reactive multistage multi component distillation; rigorous model was used to build the simulation program using MATLAB to solve MESHR equations. The validity and accuracy of the developed program model were checked with previous work for methyl acetate production. The influence of various parameters, such as number of stages, batch time and liquid hold up on the performance of batch reactive distillation was shown through the vapor and liquid profiles, temperature, reaction rate, and composition profiles.

Present Work

The present work emphasized to construct mathematical model based on Equilibrium model (EQ) for packed batch reactive distillation. The proposed model predict the composition of each component, temperature, liquid and vapor flow and reaction rate for each step time of the batch time. The following assumptions were proposed for the developed model:

1. Vapor-liquid equilibria are achieved on each segment.
2. Constant pressure drop across the column.
3. Hold-up per segment equal to liquid hold-up on segment (i.e. vapor phase molar hold-up neglected).
4. Walls of column are perfectly insulated, therefore, the heat losses neglected.
5. Reactions take place in liquid phase.
6. Each segment is at mechanical and physical equilibrium.

The performance and validity of the developed model were compared with experimental results obtained throughout the present work (23) and with the results of an existed program ASPEN PLUS (10.2) for a certain operating conditions.

Many variables have been studied that affect the purity of ETBE and the % conversion such as feed molar ratio of ethanol to tert-butanol, reflux ratios, heat duties, type of catalyst and the amount of catalyst. All physical and chemical properties are taken from literature (23).

Experimental Work

Bench experiment to produce ETBE was carried out in a batch reactor, consists of 250ml two-necked round flask one handle thermometer and the other handling stir. Heat was supplied by a jacketed heating mantle with a constant heat duty of 220 W. The reactants, ethanol and tert-butanol has been mixed with a measured amount of catalyst, and the mixture was introduced to the flask, stirred, and heated to a temperature of 75 C°, two types of catalyst have been used ($KHSO_4$, and H_2SO_4). After a certain time a sample was taken and analyzed using Gas Chromatography with porpak Q column.

The results are given in Table (1) for $KHSO_4$ catalyst using feed volume ratio of ETOH: TBA (2:1) and after 3.5hr.

Table (1): Results of $KHSO_4$ Bench experiment

x_{ETOH}	0.738
X_{TBA}	0.152
x_{ETBE}	0.092
X_{H2O}	0.018

And for H_2SO_4 catalyst using the feed volume ratio of ethanol: tert-butanol (1:1) and after 4.5hr are shown in Table (2).

Table (2): Results of H_2SO_4 Bench experiment

x_{ETOH}	0.4232
X_{TBA}	0.0724
x_{ETBE}	0.2497
X_{H2O}	0.2360

A laboratory scale packed batch reactive distillation unit has been designed to perform the present work. The unit consists of a still pot consists of a three neck-round flask connected to the distillation column through its central opening. The other two openings were connected to a thermometer to measure the bottom temperature (vapor temperature), and a line for recycling a portion of solution in the distillate.

Heat for still pot was supplied by a jacketed heating mantle; its heat duty can be controlled using an electrical controller. Heat duties are 65, 90, and 146 W. The distillation column (located above the still pot) is made of a heat resistance glass column. To avoid the heat loss, the column was insulated with rubber insulation. The height of column is 71cm height and 3.5 cm inside diameter, packed with glass rasching ring of 10 mm length, 6 mm out side diameter, and 3 mm inside diameter. At the top of column, a double pipe condenser is connected, which is used to condense the vapor leaving the top of column. Part of condensate returns back to the distillation column and the other was drained with a constant reflux ratio Figure (1).

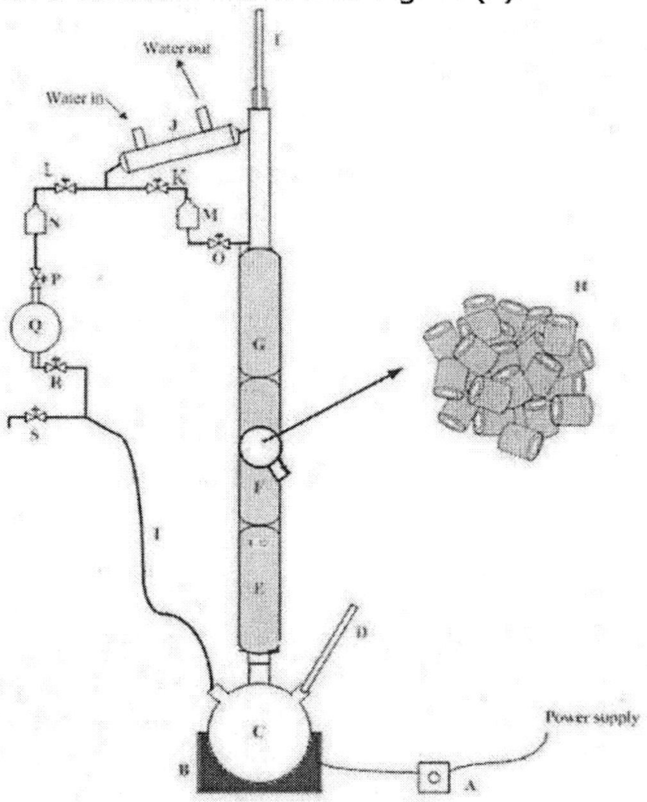

Figure 1 . Schematic Sketch of RD Unit

A: Heat duty controller; **B**: Heating mantle; **C**: Still pot; **D**: Bottom thermometer, **E**: Lower distillation column section; **F**: Intermediate distillation column section **G**: Upper distillation column section; **H**: Rasching ring; **I**: Top thermometer **J**: Condenser; **K, L, O, P**: Reflux valves; **M**: Container; **N**: Container, **Q**: Collecting vessel; **R**: Valve; **S**: Valve; **T**: Recycle line

Measured amounts of reactants were introduced to the still pot together with a desired amount of catalyst, $KHSO_4$, or H_2SO_4 (0.04 kg catalyst/1kg bulk reactant). Cooling water of condenser is started and heat to the still pot is supplied. The reaction takes place in the still between ethanol and tert-butanol then ethyl tert-butyl ether and water produced. Water have the highest boiling point among the still contents, therefore, during the distillation water accumulates in the still and little amount reaches the condenser. Samples were taken from the distillate each hour starting at the second hour because the amount of ETBE produced in the first hour is very small [4, 5, 6 and 23]. Many variables have been studied such as the feed molar ratio (FMR) of 1:1, 2:1, and 4:1, reflux ratio (Rr) of 3, 4, and 5, heat duty of (65, 90, and 146 W), catalyst amount for $KHSO_4$ catalyst, and amount of water in the feed.

Mathematical Model

Consider the batch packed reactive distillation column, the schematic model of jth segment shown in Figures (2).

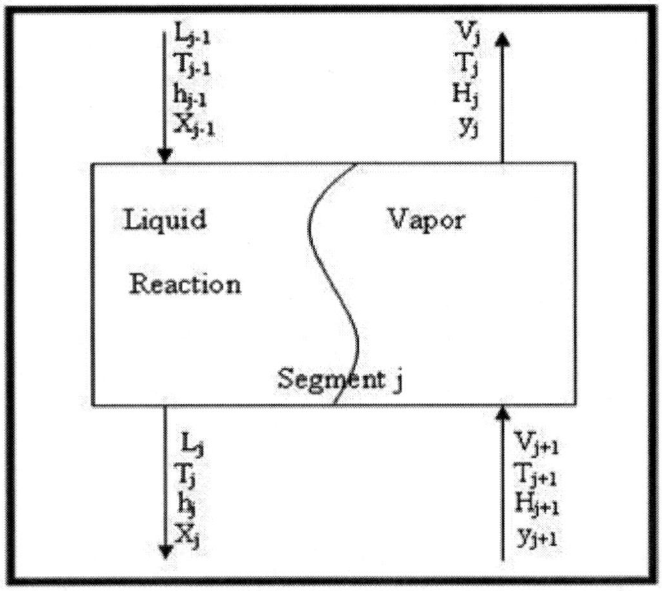

Figure 2. Schematic Diagram of equilibrium Segment

Equations that model the equilibrium segment are shown as MESHR equations. MESHR is an acronym referring to the different types of equations.

M. Total and component material balances.

$$\frac{dM_j}{dt} = V_{j+1} + L_{j-1} - V_j - L_j + R_j \qquad (1)$$

$$\frac{dM_j x_{ij}}{dt} = V_{j+1} y_{i,j+1} + L_{j-1} x_{i,j-1} - V_j y_{i,j} - L_j x_{i,j} + R_{i,j} \tag{2}$$

E. Equilibrium relation

$$y_{i,j} = K_{i,j} x_{i,j} \tag{3}$$

S. Summation equations

$$\sum_{i=1}^{c} x_{i,j} = 1, \sum_{i=1}^{c} y_{i,j} = 1 \tag{4}$$

H. Enthalpy equation

$$\frac{dM_j H_j}{dt} = V_{j+1} H_{j+1} + L_{j-1} h_{j-1} - V_j H_j - L_j h_j + R_j Hr \tag{5}$$

R. Reaction rate equations [23].

$$R = M_{cat} \frac{K_1 (a_A a_B - \frac{a_C a_D}{K_{eq}})}{1 + K_w a_D} \tag{6}$$

$$K_1 = \exp(3.55 - \frac{2286}{T}) \tag{7}$$

$$K_W = \exp(-16.16 - \frac{6636}{T}) \tag{8}$$

$$K_{eq} = \exp(-2101.93 - \frac{38672.73}{T} + 411.41 \log T - 1.592T + 1.345*10^{-3} T^2 - 5.7517*10^{-7} T^3) \tag{9}$$

The activity of i-th component was calculated using the following equation:

$$a_i = x_i \gamma_i \tag{10}$$

- Enthalpy Calculation

Enthalpy of component in liquid phase can be estimated from the heat of formation at reference temperature and the sensible heat at a desired temperature [18, 23].

$$h_i = \Delta H_{fi}^{298} + \int_{298}^{T} Cp^L dt \tag{11}$$

Evaluation of integral in Equation (11) requires knowledge of the temperature dependence of heat capacity.

$$Cp_i^L = a + bT + cT^2 + dT^{-2} \tag{12}$$

The total enthalpy of liquid phase given by:

$$h_{i,j} = \sum_{i=1}^{c} x_{i,j} h_i + H_{mix} \tag{13}$$

Where H_{mix} is the heat of mixing, for UNIFAC model heat of mixing is:

$$H_{mix} = RT \sum_{i=1}^{c} (x_i \ln \gamma_i) \tag{14}$$

The enthalpy of component in vapor phase was estimated through the integral of heat capacity in vapor phase at any temperature T with respect to a reference temperature T°:

$$H_V = \int_{T^\circ}^{T} Cp_V dt \tag{15}$$

Heat capacity of components in the vapor phase is:

$$Cp_i^V = a_V + b_V T + c_V T^2 + d_V T^{-2} \tag{16}$$

Total enthalpy of vapor phase is:

$$H_{V i,j} = y_{i,j} H_i \tag{17}$$

Heat of reaction at any temperature can be calculated from heat capacity data if the value for one temperature is known, the tabulation of data can be reduced to the completion of standard heats of formation at a single temperature [21]. The calculation of standard heats of reaction has been given by:

$$Hr^\circ = \sum_{i=1}^{c} \nu_i H_{fi} \tag{18}$$

Where the sign of stoichiometric ratio ν is positive for products and negative for reactants.

$$Hr = Hr^\circ + \int_{T^\circ}^{T} \Delta Cp_i^V dT \tag{19}$$

Enthalpy constants (a_V, b_V, c_V and d_V) for vapor and (*a*, *b*, *c* and *d*) for liquid are available in [23].

-Vapor Pressure Calculation

The vapor pressure of each component for the present system was calculated using [12, 22].

$$\ln P^{sat} = A1 + \frac{A2}{T + A3} A4 * T + A5 \ln T + A6 * T^{A7} \tag{20}$$

The constants are available in [23].

Where vapor pressure P^{sat} in Pa and T in Kelvin.

-Bubble Point Calculation

Temperatures of stages have been calculated using iterative procedure of bubble point until the summation in Equation (21) equals to one.

$$\sum_{i=1}^{m}(K_{ij}x_{ij}) = 1 \qquad (21)$$

Where K is the distribution coefficient and it can be calculated using:

$$K = \gamma_i \frac{P^{sat}}{P} \qquad (22)$$

For the present system the activity coefficient γ_i has been calculated using UNIFAC method.

-Holdup

The liquid holdup in a packed column is defined as the volume of liquid held under operating conditions per volume of packed bed. This holdup can be divided into two components, the static and the dynamic (or operating) holdup [19]. The hold up correlations is given by [20,23]:

$$M = \frac{1}{12\varepsilon}(\frac{\mu_L}{\rho_L})^{1/6}(u_L a_P)^{0.5} \qquad (23)$$

The holdup in reboiler based on the initial charge to the reboiler (M°) and it is given by [21]:

$$M_{Nt+1} = M^{\circ}_{Nt+1} - \sum_{j=0}^{Nt} M_j - \int_0^t D_t dt \qquad (24)$$

Solution of the Equilibrium Model

A computer program to solve the MESHR equations has been developed using MATLAB (R2009b) to determine the composition of components, stages temperatures, condenser and reboiler duties, liquid and vapor flow rates along stages, and reaction rate profile.

The program begins with specify all parameters consist of number of stages, reflux ratio, total pressure, feed compositions, feed flow rate, distillate rate, batch time, step time, and mass of catalyst as well as all physical properties of components. Time, and temperature loops were started, respectively over all stages. The temperature of each stage has been calculated by trail and error until the equilibrium relation applicable.

The new stages temperatures have been used in calculation of reaction rate, enthalpies of vapor, liquid and mixing. Then liquid and vapor flow rates were calculated by total material and energy balances. A tridiagonal matrix was used to find the component compositions by solving the MESHR equations, solving the matrices by eigen value, and normalizing the new compositions for each component. New sets of composition are obtained with the previous procedure for each step time of the batch time. When the compositions at different times are evaluated the program ended and the results plotted.

Results and Discussion

The validity and accuracy of the proposed unsteady state equilibrium model for production of ETBE from ethanol and tert-butanol using MATLAB (R2009b) is checked with Aspen Plus (10.2) program using a rigorous distillation model (RADFRAC) with the same operating conditions in both programs Table(3).

Results of liquid mole fraction in the distillate and bottom with time of the two programs are shown on the same plot for comparison, Figures (3) and (4). ETBE formed is mostly collected in the distillate because it has the lower boiling point 73C^{o} compared with 87, 83, and 100 C^{o} for ethanol, tert-butanol and water, respectively.

The comparison of the two programs shows good agreement in composition profiles with small deviation which is due to the lack of knowledge in the Aspen Plus data of activity coefficients, Antoine parameters, enthalpy of vapor and liquid coefficients, reaction rate kinetics and amount of catalyst used. The error between the two programs which is bounded with ± 10% lines Figure (5).

Different variables have been studied such as feed molar ratio, reflux ratio, and heat duty and the results were compared with the present experimental work.

-Feed Molar Ratio (FMR)

Experiments 1, 2, and 3 in Table (4) were performed to study the effect of changing FMR on purity of ETBE in the distillate. The observed purity of ETBE in the distillate after 3hrs is 5.619 %mol. Figure (6) shows the effect of changing FMR on the purity of ETBE, as the ratio of ethanol to tert-butanol increases the purity of ETBE decreases this is because the distillate contains the unreacted ethanol which reduces the amount of ETBE in the distillate and the best FMR obtained is 1:1.

- Effect of Changing Reflux Ratio (Rr)

Experiments 1, 2, and 3 in Table (5) were carried out to find the best reflux ratio, for the best value of FMR of 1:1. The purity of ETBE in the distillate when determining the best FMR was low, therefore, the percentage amount of water in the feed is reduced from 95% to 90% and the time increased to 6 hrs. Figures (7) to (9) show the results of liquid composition profile with time in the distillate for experimental and theoretical equilibrium model for reflux ratios of 3, 4 and 5 respectively. (Solid Lines ≡ EQ Model, Dashed Lines ≡ Experimental).

These Figures show that the purity of ETBE increases with increasing the reflux ratio and the best reflux ratio which gives higher purity of ETBE is 5. This is because when reflux ratio increases the residence time of reactants increased and the concentration of reactants in the distillate reduced therefore, the purity increased.

- Effect of Changing the Heat Duty

All the previous experiments have been carried out under heat duty of 146 W, which is the maximum operating value of the heating mantle. In order to find the effect of changing the heat duty on the purity of ETBE in the distillate; experiments with the best FMR (1:1), best reflux ratio (5), and heat duty of 65 and 90 W were performed. Figures (10) and (11) show the comparison in heat duty between experimental and theoretical model results.

Increasing the heat duty the purity of ETBE increased and the best heat duty is 146 W. When the heat duty increases the vapor rate is increased which transfers more volatile and lower boiling point component from the still to the distillate, therefore, the purity of ETBE which is the lower boiling point component in the system increases.

- Effect of Changing the Amount of Water in Feed

An experiment with 0% water in the feed for catalysts $KHSO_4$ was carried out. Figure (12) shows the comparison between experimental and theoretical model where no water used in the feed for 0.04 kg catalyst/kg reactant.

From the above Figure, the purity of ETBE increases when no water used in feed, and there is amount of tert-butanol (which is the limiting reactant) still unreacted. Higher amount of $KHSO_4$ catalyst are required to increase the purity of ETBE. Therefore, an experiment with higher amount of $KHSO_4$ has been performed.

- Effect of Changing the Catalyst Amount

$KHSO_4$ catalyst has been used with three times its amount in the previous experiments in order to increase the purity of ETBE. Figure (13) shows the liquid mole fraction profiles using feed molar ratio of 1:1, reflux ratio 5, heat duty 146W, and 0.12 kg catalyst/1kg bulk reactant.

From Figure (13) it is clear that the purity of ETBE and conversion of TBA increase with increasing the amount of $KHSO_4$. This is due to more tert-butanol and ethanol reacts to produce ETBE.

- *Effect of Changing the Catalyst Type*

Other catalyst has been chosen such as H_2SO_4 with the same amount (0.04 kg catalyst/1kg bulk reactant), for both 90% and 0% water in the feed, with the best FMR (1:1), best reflux ratio (5), and best heat duty (146W), Figures (14) and (15). The results show when sulfuric acid (H_2SO_4) is used as a catalyst the purity of ETBE is higher than that when potassium hydrogen sulfate is used. For 90% water in the feed and $KHSO_4$ catalyst, the purity of ETBE in the distillate is (0.1487) compared with (0.16615) for sulfuric acid, and for 0% water in the feed the purity is (0.29996) for potassium hydrogen sulfate compared with (0.61317) for sulfuric acid. In the experiments where H_2SO_4 is used the tert-butanol is almost consumed, while when using potassium hydrogen sulfate with the same batch time amount of tert-butanol is still unreacted even it is higher than the amount of ETBE when using 90% water and nearly equal to the amount of ETBE when no water is used for the same amount of catalyst used, this is due to that sulfuric acid is a strong acid causes the tert-butanol and ethanol to reacted faster and therefore, increases the purity of ETBE and conversion of tert-butanol, so for 0% water in the feed the conversion is (85.5775%) when using potassium hydrogen sulfate compared with (94.789%) for sulfuric acid, Figures (16) and (17).

- *Effect of Using Intermediate Reactive Section*

The reaction section has been changed to an intermediate section, where the catalyst $KHSO_4$ is placed in the middle section and the reactants have been introduced to the still, samples were taken every hour for 6h starting from the second hour. The operating conditions of the present experiment are feed molar ratio 1:1, reflux ratio 5, and boiler heat duty 146W, with 160 g amount of catalyst. Figure (18) show that the purity of ETBE in the distillate increased to (0.79407) and the conversion becomes 99.465%, when intermediate reactive section has been used. This is due to the availability of large amount of catalyst.

All above figures show a good agreement between experimental and theoretical results within $\pm 10\%$ errors. Such deviations may be attributed to the little difference in the size and shape of packings and the manual operation of the distillation column when the reflux ratio and heat duty were being set.

Conclusion

As the feed molar ratio of ethanol to tert-butanol increases, the purity of ethyl tert-butyl ether in the distillate decreases.

For the reflux ratio studied in the present work (3, 4, and 5), increasing the reflux ratio cause the purity of ETBE in the distillate to increase.

Changing the heat duty of still affect the composition where the increase of heat duty increases the purity of ETBE at the top.

For the same amount of catalyst used and the same operating conditions (FMR, reflux ratio, and heat duty). When sulfuric acid is used as a catalyst purity of ETBE in the distillate and TBA conversion is higher than that when the potassium hydrogen sulfate is used as a catalyst.

The purity of ETBE can be improved by increasing the amount of catalyst and by using intermediate reactive section, but it is also leads to augments the cost of the process.

References

1. Kenig, E. Y., Gorak, A., Pahalahti,A., Jakobsson, K., ittamaa, J. A, and Sunmacher, K. (2004) "Advanced rate-based simulation tool for reactive distillation", AIChE Journal, Vol. 50, No. 2, 322-342, February.

2. Ozbay, N., and Oktar, N. (2009) "Thermodynamic study of liquid phase synthesis of ethyl tert-butyl ether using tert-butanol and ethanol", J. Chem. Eng. Data, 54, 3208-3214.
3. Al-Arfaj, M. A. and Luyben, W. L., (2002) "Control study of ethyl tert-butyl ether reactive distillation", Ind. Eng. Chem. Res., 41, 3784-3796.
4. Assabumrungrat, S., Wangwattanasate, D., Pavarajarau, V., Praserthdam, P., Arpornwichanop, A., and Goto, S. (2004) "Production of ethyl tert-butyl ether from tert-butanol and ethanol catalyzed by β-Zeolite in reactive distillation", Korean J. Chem. Eng., 21(6), 1139-1146.
5. Umar, M., Saleemi, A. R., and Saha, B. (2009a) "Liquid phase synthesis of ETBE from ethanol and tert-butyl alcohol using macroporous and gelular ion exchange resin catalyst", The Nucleus, 46 (3), 101-107.
6. Umar, M., Saleemi, R., Feroz, N., and Qaiser, S. (2008) "Synthsis of ETBE in packed reactive distillation column using macroporous ion exchange resin catalyst", World Congress on Engineering and Computer Science (WCECS), San Francisco, USA, 22-24 October.
7. Mujtaba I. M., (2004) "Batch distillation design and operation", Series on chemical engineering, Vol. 3.
8. Matouq M., Quitain, A. T., Takahashi, K., and Goto, S. (1996) "Reactive distillation for synthesizing ethyl tert-butyl ether from low grade alcohol catalyzed by potassium hydrogen sulfate", Ind. Eng. Chem. Res., 35,982-984.
9. Sneesby, M. G., Tade, M., Datta, R., and Smith,T. N. (1997b) "ETBE synthesis via reactive distillation 1. Dynamic simulation and control aspects", Ind. Eng. Chem. Res., 36, 1870-1881.
10. Quitain, A. T., H. Itoh, and S. Goto (1999b) "Reactive distillation for synthesizing ethyl tert-butyl ether from bioethanol", Journal of Chemical Engineering of Japan, Vol. 3, No. 3, 280-287.
11. Bisowarno B. H., and Tade, M. O. (2000) "Dynamic simulation of startup in ethyl tert-butyl ether reactive distillation with input multiplicity", Ind. Eng. Chem. Res., 39, 1950-1954.
12. Yang, B., Yang, S., and Wang, H. (2001) "Simulation for the reactive distillation process to synthesize ethyl tert-butyl ether", J. Chem. Eng. of Japan, 34, 1165-1170.
13. Jhon, Y. H., and Lee, T. (2003) "Dynamic simulation for reactive distillation with ETBE synthesis", Separation and Purification Technology, 31, 301-317.
14. Varsili, D., and Dogn, T. (2005) "Simultaneous production of tert-amyl ethyl ether and tert-amyl alcohol from isoamylene-ethanol-water mixtures in a batch reactive distillation column", Ind. Eng. Chem. Res., 44, 5227-5232.
15. Hamza, T. M., and Haweel, C. K. (2005) "Study of performance of batch reactive distillation", M. Sc. Thesis, University of Baghdad, Baghdad, Iraq.
16. Umar, M., Patel, D., and Saha, B. (2009b) "Kinetic studies of liquid phase ethyl tert-butyl ether (ETBE) synthesis using macroporous and gelular ion exchange resin catalyst", Chem. Eng. Sci., 46, 4424-4432.
17. Nakkash, N. B., Al-Habobi, N., and Hamodee, H. M. (2010) "Simulation of batch reactive distillation for methyl acetate production", Spring National Meeting AIChE, T8000. Advances in distillation and absorption, 23-March.
18. Perry, R. H., and Green, D. W. (1997) "Perry's chemical engineer's handbook", McGraw-Hill Companies, Inc.
19. Wagner, I., Stichlmair, J., and Fair, J. R. (1997) "Mass transfer in beds of modern, high efficiency random packings", Ind. Eng. Chem. Res., 36, 227-237.
20. Kister, H. Z. (1992)"Distillation design", McGraw-Hill, Inc., New York .
21. Seader, J.D., and Henley, E. J. (1998) "Separation process principles", John Wiley and Sons, Inc., New York.

22. Walas, S. M. (1985) "Phase equilibria in chemical engineering", Butterworth Publications, London.
23. Al-Janabi, N. K. (2010) "Simulation and Experimental Investigation to produce Ethyl Tert-Butyl Ether using Reactive Distillation", M.Sc. Thesis, Al-Nahrain University, Baghdad, Iraq.

Notations

Symbols	Notations
A1,A2,A3,A4 ,A5	Antoine Equation Constants
a_i	Activity
A, b, c, d	Constants of Heat Capacity of Liquid Phase
a_v, b_v, c_v, d_v	Constants of Heat Capacity of Vapor Phase
Cp	Specific Heat
E	Energy
H_v	Enthalpy of Vapor Phase
H	Enthalpy of Liquid Phase
H_{mix}	Heat of Mixture
Hr°	Reference Heat of Reaction
ΔH_{vb}	Heat of Vaporization at Normal Boiling Point
Hr	Heat of Reaction
h^V	Heat Transfer Coefficient in Vapor Phase
h^L	Heat Transfer Coefficient in Liquid Phase
K	Equilibrium Constant
L	Liquid Flow Rate
M_{cat}	Mass of Catalyst
M	Liquid Hold-Up
N	Total Number of Component
P	Pressure
P^{sat}	Saturated Pressure
R_i	Reaction Rate
T^o	Reference Temperature
V	Vapor Flow Rate
X	Liquid Mole Fraction
Y	Vapor Mole Fraction

Greek Letters

γ	Activity Coefficient

Subscripts

i, j	Component indices
L	Liquid
V	Vapor
N_t	Total Number of Stages
N_{t+1}	Reboiler

Superscripts

Sat	Saturated
L	Liquid
V	Vapor

Abbreviations

RD	Reactive Distillation
ETBE	Ethyl Tert-Butyl Ether
MTBE	Methyl Tert-Butyl Ether
FMR	Feed Molar Ratio
IB	Isobutene
ETOH	Ethanol
TBA	Tert-Butanol
FVR	Feed Volume Ratio
Rr	Reflux Ratio

Table 3. Operating Conditions for Comparison of Proposed EQ Program and Aspen Plus [4].

Pressure (Pa)	101325
Feed molar ratio (TBA:ETOH:H_2O)	1:1:38
Composition [% mol] ETOH TBA H_2O	 2.5 2.5 95
Feed flow rate (mol/s)	$2.7*10^{-3}$
Reflux ratio	1.5
Catalyst weight per stage (kg)	0.04
Total stages	16
Rectification stages	5
Reaction stages	6
Stripping stages	5
Feed stage	12
Heat duty (W)	26.3

Table 4. Studied Operating Conditions for Changing FMR

No.	Reflux ratio	FMR	Heat duty	Initial water content	Catalyst used
1	Total	1:1	146 W	95%	$KHSO_4$
2	Total	2:1	146 W	95%	$KHSO_4$
3	Total	4:1	146 W	95%	$KHSO_4$

Table 5. Studied Operating Conditions for the Changing the Reflux Ratio

No	Reflux ratio	FMR	Heat duty	Initial water content	Catalyst used
1	5	1:1	146 W	90%	$KHSO_4$
2	4	1:1	146 W	90%	$KHSO_4$
3	3	1:1	146 W	90%	$KHSO_4$

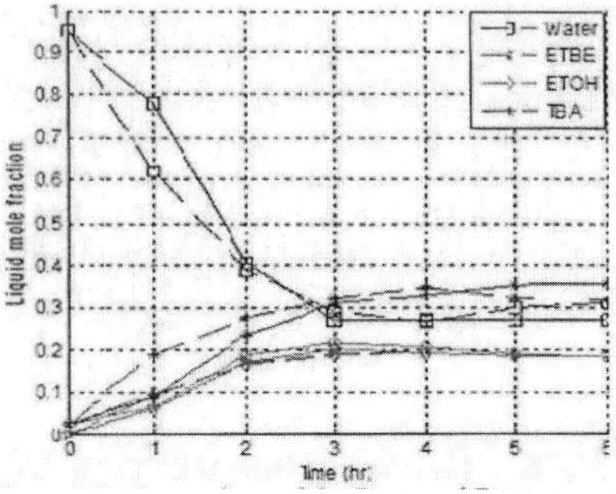

Figure 3. Comparison of the Proposed Program and Aspen Plus, Distillate, Solid lines: ASPEN PLUS, Dashed Lines: Proposed Program

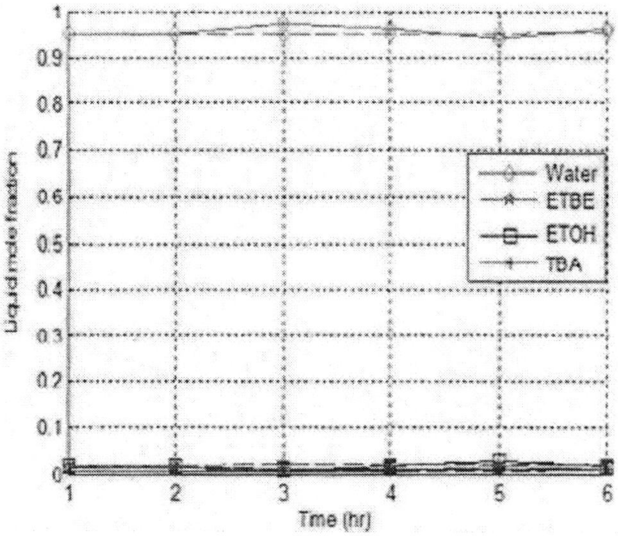

Figure 4. Comparison of the Proposed Program and Aspen Plus, Bottom, Solid lines: ASPEN PLUS, Dashed Lines: Proposed Program

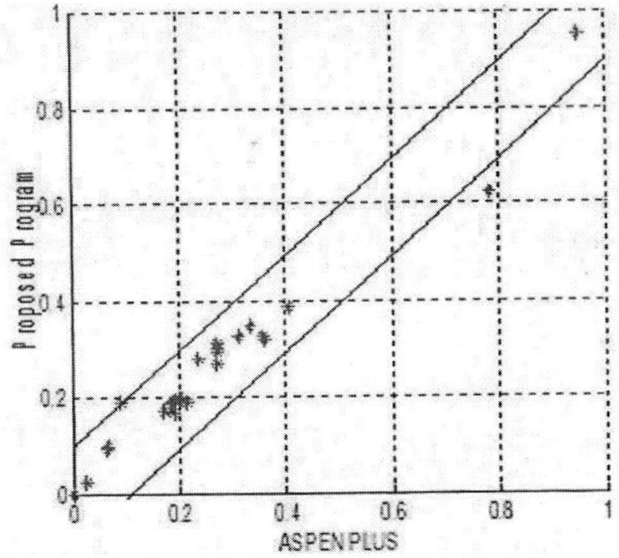

Figure 5. Error between the Two Programs in the Distillate

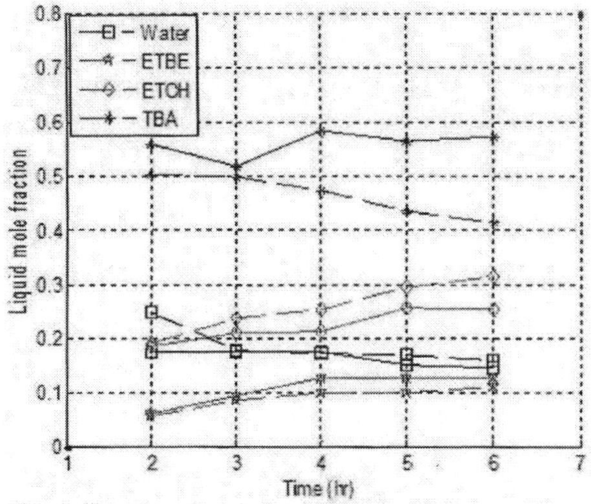

Figure 7. Experimental and EQ Model Composition profile in the Distillate for Rr=3, Heat Duty=146W

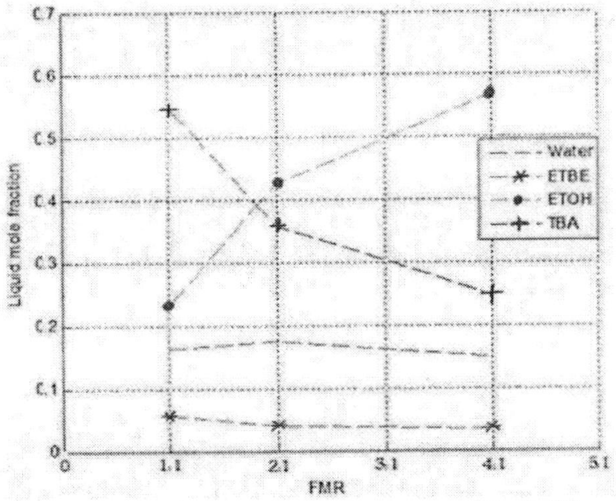

Figure 6. Liquid Mole Fractions with Different FMR

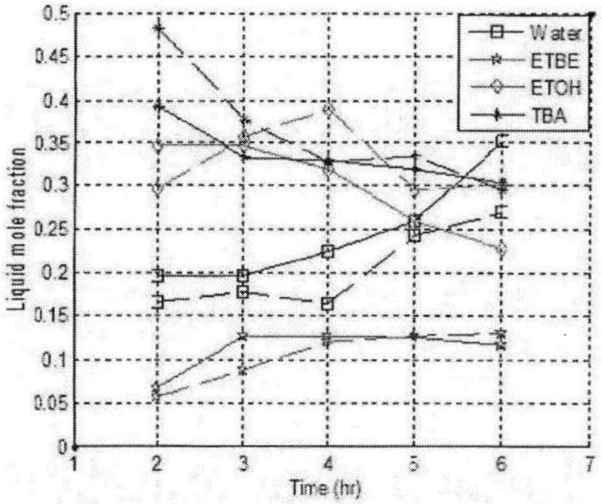

Figure 8. Solid Lines≡EQ, Dashed Lines≡Experimental, Rr=4, Heat Duty=146W

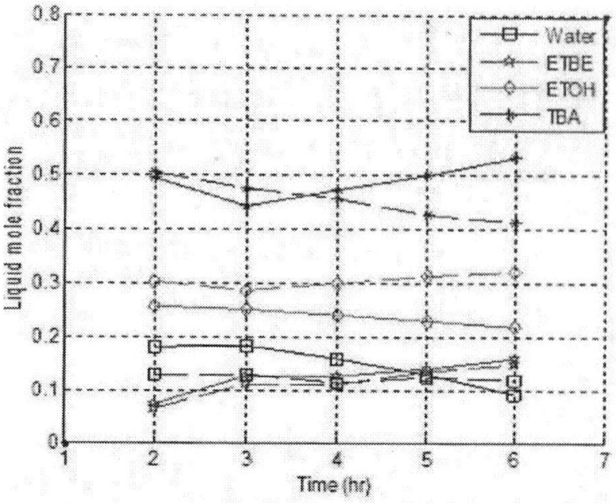

Figure 9. Solid Lines≡EQ, Dashed Lines≡Experimental, Rr=5, Heat Duty=146W

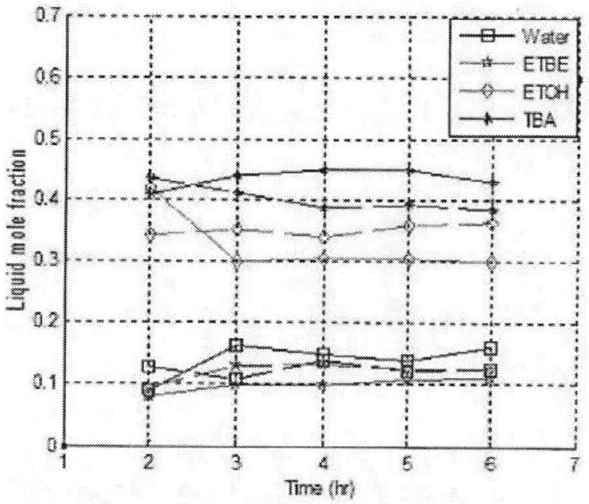

Figure 10. Solid Lines≡EQ, Dashed Lines≡Experimental, Heat Duty=65W

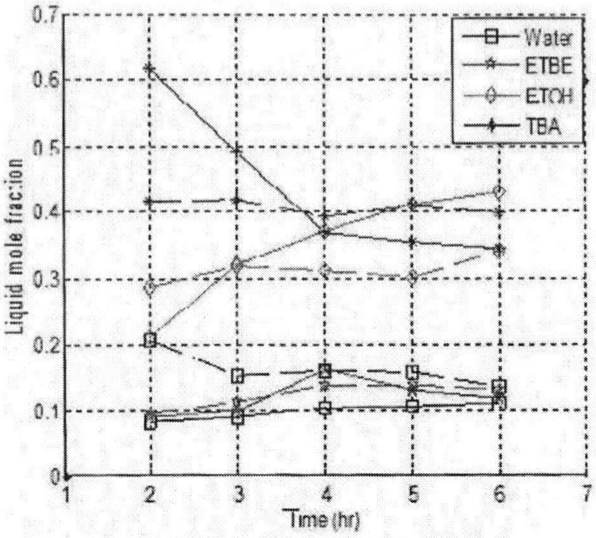

Figure 11. Solid Lines≡EQ, Dashed Lines≡Experimental, Heat Duty=90W

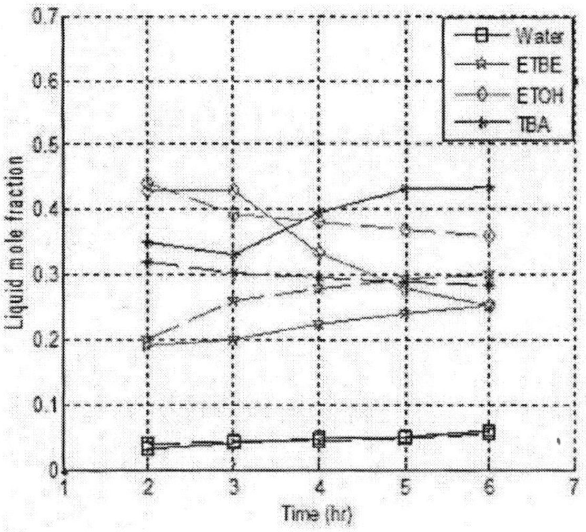

Figure 12. Solid Lines≡EQ, Dashed Lines≡Experimental, 0%Water in Feed

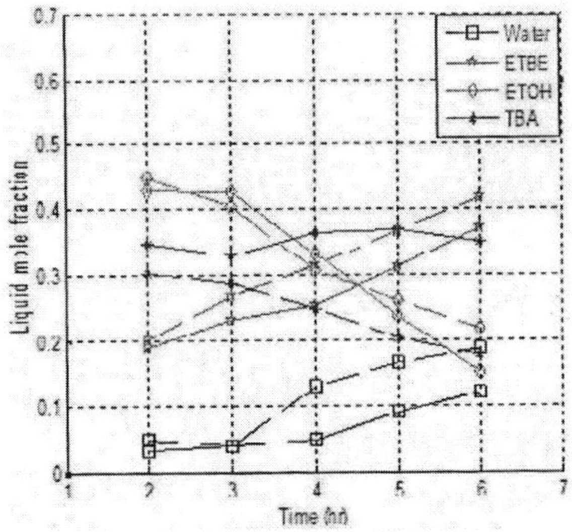

Figure 13. Solid Lines ≡ EQ, Dashed Lines ≡ Experimental, 0%Water in Feed with Three Times Catalyst $KHSO_4$

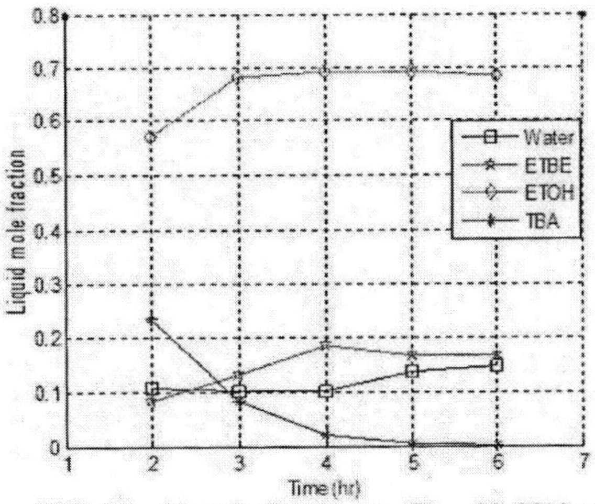

Figure 14. Liquid mole fraction profile with 90% water and catalyst H_2SO_4

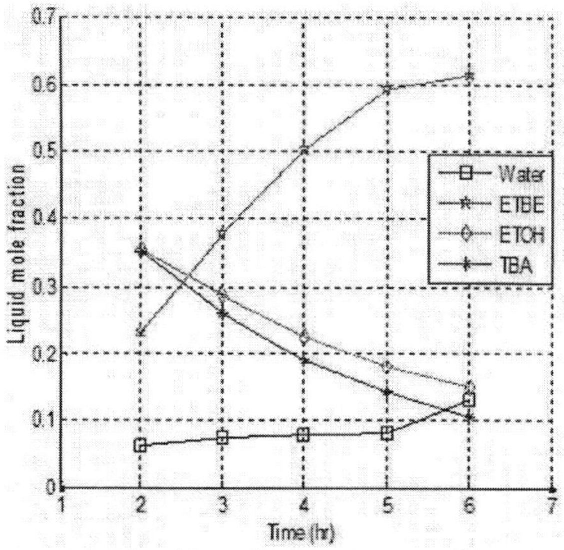

Figure 15. Liquid mole fraction profile with 0% water and catalyst H_2SO_4

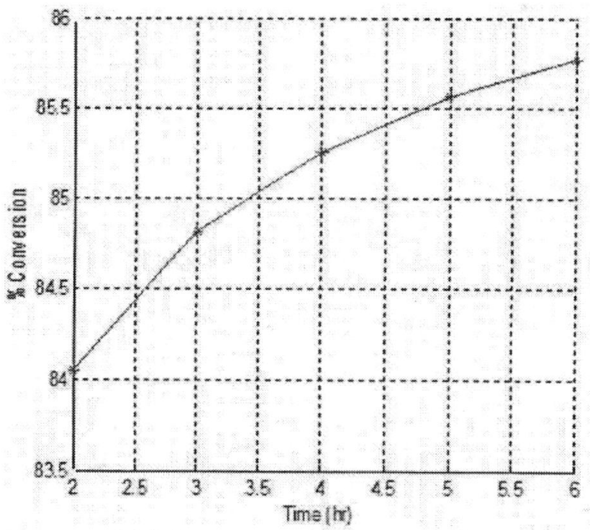

Figure 16. Conversion profile for 0% water and catalyst $KHSO_4$

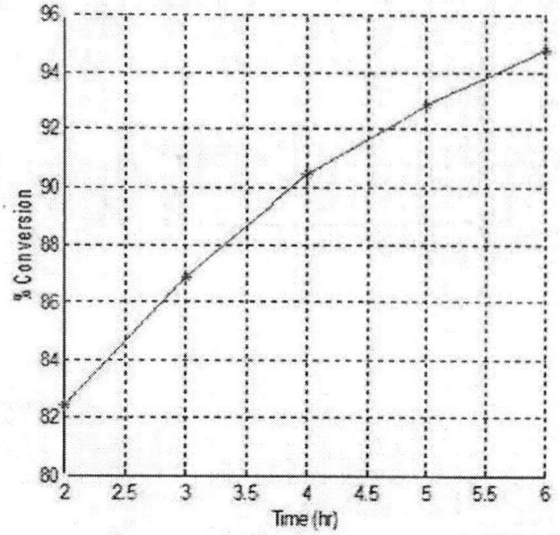

Figure 17. Conversion profile for 0% water and catalyst H_2SO_4

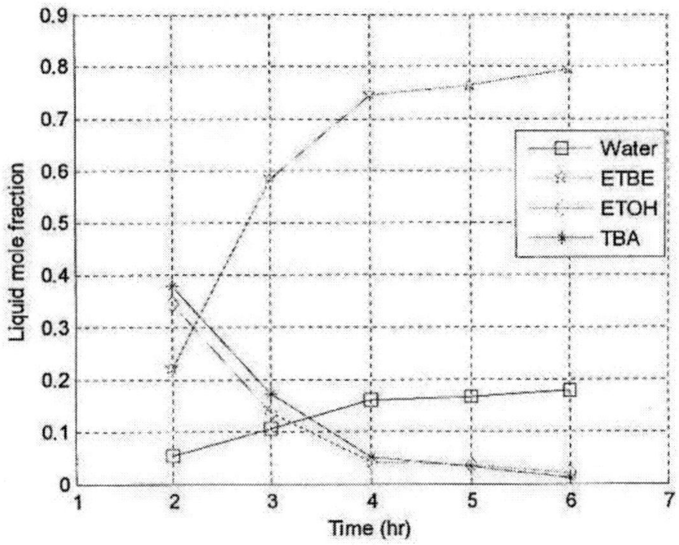

Figure 18: Liquid mole fraction profile using intermediate reactive section

SIMULATION OF CONTINOUS REACTIVE DISTILLATION OF ETBE PRODUCTION FROM ISOBULENE

Nada Bahjat Nakkash, Nadeen Khalid Al-Janabi
Al-Nahrain University, Baghdad – Iraq

Abstract

The present work concerned with simulating a continuous multicomponent reactive distillation (RD) using equilibrium model for the production of ethyl tert-butyl ether from isobutene and ethanol. A simulation model was performed for RD column with ETBE synthesis using MATLAB (R2009b) to solve MESHR equations.

The developed simulation model was checked and compared with a previous work and RADFRAC model of ASPEN PLUS 10.2, the results gave a good agreement and with $\pm 10\%$ error.

Split feed condition was considered and the influence of various parameters on the performance and behavior of RD for ETBE production, such as reflux ratios in range of 2 to 8, mass of catalyst of 60 to 190 g, number of reaction stages of 3 and 4 has been studied.

The best design conditions for operating RD distillation column to obtain high purity of ETBE 89.89% with 40.7% selectivity and 99.82% conversion was at reflux ratio 5, number of stages 10, and mass of catalyst 120g per stage using single feed. When split feed is considered with ETBE fed at stage 2 and ETOH, IB and n-B at stage 6 and when ETBE and ETOH fed at stage 2, IB and n-B at stage 6 the best design conditions was at reflux ratio 5, number of stages 10, mass of catalyst 90g per stage and three reactive stages.

Keywords: *Continuous Reactive Distillation, Simulation, ETBE Production*

Introduction

Reactive distillation involves combined simultaneous chemical reaction and distillation. The chemical reaction usually takes place in the liquid phase or at the surface of a solid catalyst in contact with the liquid phase [1]. The use of reactive distillation has grown because it results in less expensive and more efficient processes for some chemical systems. There is increasing interest in the use of ethyl tert-butyl ether (ETBE) for gasoline blending as a replacement for methyl tert-butyl ether (MTBE) because of the latter's environmental problems [2].

The basic reaction combines ethanol and isobutene to form ETBE.

$$iC_4^{=} + Ethanol \Leftrightarrow ETBE$$

The isobutene feed stream contains both isobutene and normal butene, and the latter does not participate in the reaction. The chemically inert normal butene is much lighter than ETBE, so the distillate is mostly n-butene and the bottom is mostly ETBE. Any unreacted isobutene leaves mostly in the distillate. If an excess of ethanol is fed to the column, this ethanol will typically leave in the bottoms and be blended with ETBE in gasoline pool [2, 3].

Sneesby et al. [3] presented continuous mathematical models for reactive distillation of ETBE synthesis from isobutene and ethanol using both Pro II and Speed Up simulation software. Comparison between the two programs shows excellent agreement with each other.

Quitain et al. [4] proposed an industrial scale process for synthesis of ETBE from tert-butanol and simulated using ASPEN PLUS. The overall conversion of TBA and the overall ETBE selectivity are 98.9% and 99.9% respectively.

Al-Arfaj and Luyben[5] explored the effect of number of trays in the rectifying and stripping sections of reactive distillation columns for ETBE and a hypothetical system. This work demonstrates that adding additional trays in a reactive distillation column does not degrade performance, provided the specified degrees of freedom are appropriately chosen.

Tade and Tian [6] showed that the conversion could be inferred from multiple process temperatures. For 10-stages ETBE reactive distillation pilot plant process with six temperature measurements, a third-order nonlinear inferential model was developed using MATLAB software to infer the reactant conversion with sufficiently high accuracy. The two temperatures employed in the model were the bottom reactive section temperature and the reboiler temperature, respectively.

Al-Arfaj and Luyben [2] explored two process configurations for ETBE synthesis: a design with two fresh reactant feed streams and a design with a single mixed reactant feed. Several control structures are investigated. Their results showed that the double feed system requires internal composition control to balance the stoichiometry along with temperature control to maintain product purity, while, the single feed case is effectively controlled with only a temperature controller.

Jhon et al., [7] presented a dynamic simulation model using MATLAB software for reactive distillation column of 20 stages with ETBE synthesis. A structured and simple iterative algorithm was devised to estimate the vapor flows with the fast convergence under the rigorous energy balance. Both total and partial condensers are considered, and the reactive distillation performance for these condensers was observed. By comparison of their influence on purity of ETBE, total condenser is the better one.

Yanghong et al. [8] studied the variation of reflux ratio, reboiler rate, and feed molar ratio of ETOH/IB and their influence on the conversion of IB on ETBE purity. Experimental results showed that higher conversion of IB was obtained with feed molar ratio of 1 to 1.1, reflux ratio 6, and 0.115 m^3/hr reboiler rates. For these conditions conversion was exceeded 90%.

Bisowarno et al. [9] presented a mathematical design model of single and double feed reactive distillation column for ETBE production and investigated the effects of separation and reaction stages on the overall performance. Their results showed that for the single feed column, longer column requires both reflux ratio and reboiler duty to be adjusted to optimize the ETBE purity, while the increasing of reflux ratio does not reduce the ETBE purity for the shorter column. For the double feed column, increasing reflux ratio does not degrade the column performance.

Bolun et al. [10] modeled a process for synthesis of ETBE with ASPEN PLUS simulation package. They were observed the input multiplicity for a range of reboiler duty of several values of reflux ratio. These results can be used to avoid the excessive energy consumption and achieve optimum design of reactive distillation column. They found that the best reflux ratio is 5 over a narrow range of reboiler duty (78-83 kW).

Gomez et al. [11] proposed a mixed integer nonlinear programming (MINLP) formulation for optimal design of a reactive distillation column based on a generic nonequilibrium model. Catalytic distillation of ethyl tert-butyl ether production from isobutene and ethanol is considered as an illustrative example.

Mohammed et al. [12] developed a steady state model for ETBE and MTBE synthesis using MATLAB program. Liquid composition, vapor composition, and temperature profiles were determined. Also, the residue curve maps were plotted for reacted system and system without reaction.

Present Work

The present work emphasized to construct a simulation model for continuous equilibrium reactive distillation of ethyl tert-butyl ether (ETBE) production from isobutene and ethanol over an acidic ion exchange resin, Ambelyst 15.

The profiles of stage temperature, vapor and liquid flow rates, rate of reaction, composition of vapor and liquid were computed using MESHR equations.

The percentage conversion and selectivity are calculated for both single feed and split feed conditions using different amounts of catalyst of 60g to 190g, reflux ratios of 2 to 8 and number of reaction stages 3 and 4.

The initial input conditions and data for the developed simulation model of RD are given in Table (1) for single intermediate feed and the simulation results accuracy were checked with ASPEN PLUS 10.2 software[13] and with previous work [3].

The physical and chemical properties are taken from literature [14, 15] and due to the highly nonideal nature of ether-alcohol-alkenes system. UNIFAC model has been used to predict liquid phase activity coefficient.

In the present work the vapor phase is assumed to be nearly ideal, liquid feed is used, total condenser to give a maximum heat removed at the top of the column, constant molar liquid holdup, neglect the vapor holdup, perfect mixing at each stage.

Mathematical Model

The reactive distillation column was modeled using N-equilibrium stages including total condenser and partial reboiler. Schematic of RD column is shown in Figure (1), and schematic of single stage is in Figure (2).

The following assumptions were made when formulating the reactive distillation process:

1. The vapor and liquid are in equilibrium at each stage.
2. The reaction occurs in the liquid phase.
3. The heat loss and mass transfer resistance can be neglected.

MESH equations and reaction rate were used as the basic equations for the steady state reactive distillation model.

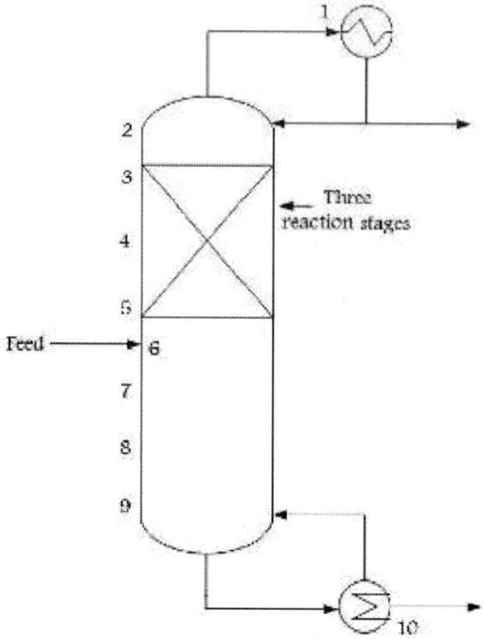

Figure 1. Schematic of RD column

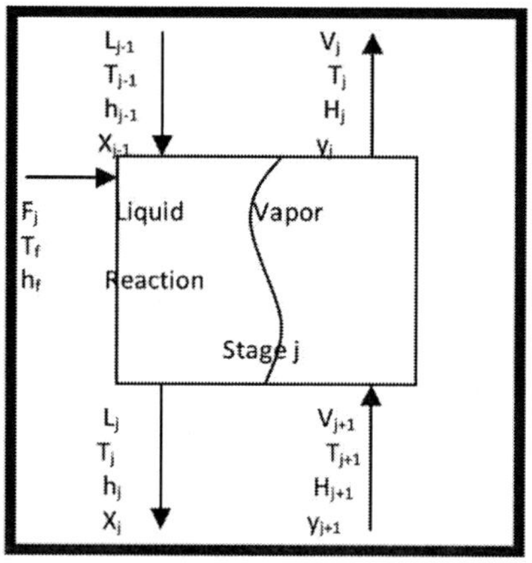

Figure 2. Schematic of a single stage

The MESHR equations are:

-Separation Stage

Total material balance:

$$L_{j-1}+V_{j+1}-V_j-L_j=0 \tag{1}$$

Component balance:

$$V_{j+1}y_{i,j+1}+L_{j-1}X_{i,j-1}-V_jy_{i,j}-L_jX_{i,j}=0 \tag{2}$$

Energy balance:

$$V_{j+1}H_{j+1}+L_{j-1}h_{j-1}-V_jH_j-L_jh_j=0 \tag{3}$$

Phase equilibrium relation:

$$y_{i,j}=K_{i,j}x_{i,j} \tag{4}$$

Summation equations:

$$\sum_{i=1}^{c}X_{i,j}=1,\sum_{i=1}^{c}y_{i,j}=1 \tag{5}$$

-Reactive Stage

Total material balance:

$$L_{j-1}+V_{j+1}-V_j-L_j+R_j=0 \tag{6}$$

Component balance:

$$V_{j+1}y_{i,j+1}+L_{j-1}X_{i,j-1}-V_jy_{i,j}-L_jX_{i,j}+r_{ij}=0 \tag{7}$$

Energy balance:

$$V_{j+1}H_{j+1}+L_{j-1}h_{j-1}-V_jH_j-L_jh_j-r_j\Delta Hr=0 \tag{8}$$

-Total Condenser

Total material balance:

$$V_2-L_1=0 \tag{9}$$

Component balance:

$$V_2y_{i,2}-L_1X_{i,1}=0 \tag{10}$$

Energy balance:

$$V_2H_2-L_1h_1=0 \tag{11}$$

-Partial Reboiler

Total material balance:

$$L_{Nt-1}-V_{Nt}-L_{Nt}=0 \tag{12}$$

Component balance:

$$L_{Nt-1}X_{i,Nt-1}-V_{Nt}y_{i,Nt}-L_{Nt}X_{i,Nt}=0 \tag{13}$$

Energy balance:

$$L_{Nt-1}h_{Nt-1}-V_{Nt}H_{Nt}-L_{Nt}h_{Nt}=0 \tag{14}$$

-The Reaction Kinetics

ETBE is produced by the reversible reaction of isobutene and ethanol over an acid catalyst, such as acidic ion-exchange resin, Amberlyst 15.

$$(CH_3)_2C=CH_2+C_2H_5OH \Leftrightarrow (CH_3)_2COC_2H_5$$

The reaction rate expression [2, 3, 5] is as follows:

$$k_{ETBE} = 10.387 + \frac{4060.59}{T} - 2.89055 \ln T$$

$$- 0.0191544\, T + 5.28586 * 10^{-5} T^2 - 5.32977 * 10^{-8} T^3 \quad (15)$$

Generalized rate equation

$$r = \frac{m_{cat} K_{rate} a_{ETOH}^2 (a_{IB} - \frac{a_{ETBE}}{K_{ETBE} a_{ETOH}})}{(1 + K_A a_{ETOH})^3} \quad (16)$$

$$\ln K_A = -1.0707 + \frac{1323.1}{T} \quad (17)$$

$$K_{rate} = 7.418 * 10^{12} \exp(\frac{-60.4}{RT}) \quad (18)$$

ETBE reaction system can also include a side reaction; the dimerization of isobutene to produce di-isobutene:

$$(CH_3)_2C{=}CH_2 + (CH_3)_2C{=}CH_2 \Leftrightarrow [(CH_3)_2C{=}CH_2]_2$$

However, this side reaction is eliminated when ethanol is in excess over the isobutene [2].

If any water is present in the reaction environment, one undesirable reaction (dehydration of isobutene to form isobutyl alcohol) is possible.

$$(CH_3)_2C = CH_2 + H_2O \Leftrightarrow (CH_3)_3COH$$

A computer program to solve the MESHR equations has been developed using MATLAB (R2009b) to determine the composition of vapor and liquid, stages temperatures, liquid and vapor flow rates through the tower, reaction rate at each stage, and condenser and reboiler duties.

The program begins with specify all specifications consist of number of stages, reflux ratio, total pressure, feed compositions, feed flow rate, distillate rate, and mass of catalyst as well as all physical properties of components. Iteration of stage temperature and composition loops was started respectively on each stage. The temperature of each stage has been calculated by trial and error until the equilibrium relation applicable, where the distribution coefficient computed from $K_{i,j} = \gamma_i \frac{P^S}{P}$ for nonideal system. Antoine equation has been used to predict the vapor pressures P^S and the coefficients are taken from literature [14].

$$\ln\frac{P^S}{P_C} = (1-w)^{-1}(Aw + Bw^{1.5} + Cw^3 + Dw^6) \quad (19)$$

Where $w = 1 - \frac{T}{T_C}$ (20)

The liquid activity coefficient γ_i for the reaction mixture of ETBE synthesis can be computed using UNIFAC method [15]. The equations and parameters to evaluate γ_i for ETBE synthesis are taken from literature [2, 15]. The new stages temperatures have been used in calculation of reaction rate and liquid and vapor flow rates by material and energy balances. For energy balance, total liquid enthalpy H_{Lj} and vapor enthalpy H_{Vj} are calculated and their coefficients were obtained from literature [14].

The vapor enthalpy

$$H_i(j) = h_{fi} + C_{1iv}(T_i(j) - T_o) + C_{2iv}(T_i(j)^2 - T_o^2)/2 + C_{3iv}(T_i(j)^3 - T_o^3)/3 + C_{4iv}(T_i(j)^4 - T_o^4)/4 \quad (21)$$

$$H(j) = \Sigma H_{Vi}(j) * y_{i,j} \quad (22)$$

The liquid enthalpy

$$h_i(j) = c_{1il}(T_i(j) - T_o) + c_{2il}(T_i(j)^2 - T_o^2)/2 + c_{3il}(T_i(j)^3 - T_o^3)/3 + c_{4il}(T_i(j)^4 - T_o^4)/4 \quad (23)$$

$$H_{mix} = RT\, \Sigma\, x_i \ln\gamma_i \quad (24)$$

$$h(j) = \Sigma\, h_i(j) * x_{i,j} + H_{mix} \quad (25)$$

A tridiagonal matrix was used to find the compositions of each component by solving the MESHR equations, solving the matrices by Eigen value, and normalizing the compositions, new values of temperatures, reaction rates, vapor and liquid flow rates on each tray have been calculated. Then the condenser and reboiler heat duties were computed.

Results and Discussions

The proposed model is used to calculate the temperature, liquid and vapor flow rates, reaction rate and composition profiles and the heat input to the still and the heat removed from the condenser for ETBE production from Isobutene and Ethanol. The results of the developed model using initial conditions given in Table (1) are shown in Figure (3). Figure (3) shows a good agreement in composition profile of the developed program and Aspen Plus results with about $\pm$10%error, as in Figure (4).

Deviation of the proposed program and Aspen Plus is expected because Aspen Plus uses built in data such as the data of UNIFAC method, enthalpy, reaction rate equation including the mass of catalyst, and Antoine equation.

The purity of ETBE at the bottom of tower is 89.89% with40.7% selectivity and 99.82% conversion when single feed is used at stage 6, reflux ratio 5, amount of catalyst 120g per stage and three reaction stages, Figure(3). As the amount of catalyst decreased to 110g per stage the purity decreases to 86.1% with 99.75 % conversion and 38.97 % selectivity.

When split feed is considered for the same operating conditions, ETBE fed to stage 2 and IB, ETOH and n-B fed to stage 6 using 100g per stage catalyst. ETBE purity increased to 99.69%, Figures (5). When ETBE and ETOH fed to stage 2 and IB and n-B fed to stage 6 the purity of ETBE, conversion and selectivity stays approximately the same. Thus it is obvious that using split feed the purity of ETBE, conversion and selectivity increases with decreasing the amount of catalyst used.

When the number of stages increased to 4 and the mass of catalyst increased to 190g per stage, the purity of ETBE decreased to 94.85% when ETBE fed to stage 2 and ETOH, IB and n-B fed to stage 6 Figure(6). While ETBE purity becomes 96.33% when ETBE and ETOH fed to stage 2 and IB, and n-B fed to stage 6 Figure(7). Thus, when the reactive stages increased from 3 to 4 the purity of ETBE decreases, this is because increasing the reactive stages increasing ETBE in the lower stages and shifts the chemical equilibrium back to the reactants will decrease the purity of ETBE.

Figure (8, 9, 10, and 11) show the effect of change reflux ratio on conversion and Figures (12, 13, 14 and, 15) show the effect of changing the reflux ratio on selectivity for split feed condition using three and four reactive stages. In all cases increases the reflux ratio caused the conversion to increase where the best reflux ratio is 5, increasing the reflux above the best the conversion remains nearly constant. Also increasing the reflux ratio will increase the selectivity of ETBE, for all cases the best reflux ratio with respect to selectivity is 7. Although increasing the reflux ratio over 5 (which is the best value for conversion) increases the selectivity by nearly 10%, in practical applications operating at higher reflux ratio is not economic due to the requirement of higher heat duty. When the reflux ratio increases the residence time of reactants increased and the concentration of reactants in the distillate decreased and therefore, the conversion and selectivity increased.

For reflux ratio 5 and split feed condition as the mass of catalyst increases there is a slight increase in purity of ETBE, Tables (2, 3, 4, and 5).

Conclusion

1. The simulation model developed gives a good agreement with ASPEN PLUS 10.2 program with about $\pm 10\%$ errors for the same operating conditions
2. Using single feed the purity of ETBE in the bottom is 89.89% for 120g/stage of catalyst, when split feed is used the purity of ETBE increased to 99.69% for 100g/stage catalyst.
3. Increasing the number of reactive stages above three increases ETBE in the lower stages and shifts the chemical equilibrium back to the reactants and therefore decreases the purity of ETBE.
4. Increasing the reflux ratio increase the residence time of reactants and decrease the concentration of reactants in the distillate. Thus, the purity of ETBE increases and the best reflux ratio is 5.
6. Increase the mass of catalyst for reflux ratio 5 and split feed condition cause slight increase in the purity of ETBE.

References

1. Seader J. D. and Henley J. R. (1998), "Separation process principles"; John Wiley and Sons.
2. Al-Arfaj M. A. and Luyben W. L. (2002), "Control study of ethyl tert-butyl ether reactive distillation"; Ind. Eng. Chem. Res.; 41, 3784-3796.
3. Sneesby M. G., Tade M. O., Datta R., and Smith T. N. (1997); "ETBE synthesis via reactive distillation. 1. Steady state simulation and design aspects"; Ind. Eng. Chem. Res.; 36, 1855-1869.
4. Quitain A., Itoh H., and Goto S.(1999); "Industrial scale simulation of proposed process for synthesizing ethyl tert-butyl ether from bioethanol" Journal of chemical engineering of Japan, vol. 32, No. 4, pp. 539-543.
5. Al-Arfaj M. A. and Luyben W. L.(2000); "Effect of number of fractionating trays on reactive distillation performance", AIChE Journal, December, vol. 46, No. 12, 2417-2425,.
6. Tade M. O., and Tian Y.(2000); "Conversion influence for ETBE reactive distillation", Separation and purification technology, 19, 85-91.
7. Jhon Y. H., and Lee T.(2003); "Dynamic simulation for reactive distillation with ETBE synthesis", Separation and purification technology, 31, 301-317.
8. Yanghong Li, Xigang Yuan, Shaomin Wa (2003); "Zeolite film packing application to synthesis of ETBE in a catalytic distillation column, Fuel Chemistry Division Preprints,48,(1),439.
9. Bisowarno B. H., Tade M. O., and Tian Y. (2004); "Interaction of separation and reactive stages on ETBE reactive distillation", AIChE Journal, March, vol. 50, No. 3, 646-653.
10. Bolun Y., Jiang W., Guosheng Z., Huajum W., and Shiqing L. (2006) "Multiplicity analysis in reactive distillation column using ASPEN PLUS", Chinese J. Chem. Eng., 14, 3, 301-308.
11. Gomez J. M., Reneaume J., Roque M., Meyer M., and Meyer X. (2006) " A mixed integer non linear programming formulation for optimal design of a catalytic distillation column based on a generic non equilibrium model", Ind. Eng. Chem. Res.; 45, 1373-1388.
12. Mohammed M. Z., and Shakoor Z. M. (2009)" modeling and simulation of

reactive distillation with MTBE synthesis" M. Sc. Thesis, University of Technology, Baghdad-Iraq.

13. ASPEN Technology Inc. (1993) "The Speed UP User's Manual"; Aspen Technology, Inc: Cambridge, MA.

14. Reid R.C.; Prausnitz,J.M.; Poling B.E. (1987); "The properties of Gases and Liquids", 4th Edt., Mc-Graw Hill; New York.

15. Smith J.M., Vanness, H.C., Abbot M. M. (2001) "Introduction to Chemical Engineering Thermodynamics" sixth edition. Mc-Graw Hill, New York.

Nomenclature

Aw, Bw, Cw, Dw	Antoine coefficients	
A	Activity of species	
c1, c2, c3, c4	Coefficients of liquid enthalpy	
C1, C2, C3, C4	Coefficients of vapor enthalpy	
H	Enthalpy of vapor	kJ/kmol
H_{mix}	Enthalpy of mixing	kJ/kmol
Hr	Heat of reaction	kJ/kmol
H	Enthalpy of liquid	kJ/kmol
K	Distribution coefficient	
K_A	Absorption equilibrium constant	
K_{rate}	Reaction rate constant	mol/hr. g of catalyst
K_{ETBE}	ETBE reaction equilibrium constant	
L	Liquid flow rate	kmol/hr
m_{cat}	Mass of catalyst	G
P	Total pressure	Atm
P^S	Vapor pressure	Bar
P_C	Critical pressure	Bar
R	Gas constant=8.314	J/mol. K
R	Reaction rate	mol/hr
T	Temperature	K
T_o	Reference temperature	K
T_C	Critical temperature	K
V	Vapor flow rate	kmol/hr
X	Liquid mole fraction	
Y	Vapor mole fraction	

Greek letters

Γ	Activity coefficient
Δ	Change of difference

Subscripts

I	Referring to species
J	Referring to stage
1	Condenser,stage1
2	Stage 2
Nt	Reboiler

Table 1. Initial input simulation data [3]

Feed composition	ethanol=0.091; isobutene=0.073; ETBE=0.291; n-butene=0.545;
Total number of stages	10
Reflux ratio	5
Total pressure	9.5 atm
Condenser	Total
Reboiler	Partial
Number of Reaction stages	3
Feed stage location	6
Feed rate	0.407 kmole / hr
Distillate rate	0.217 kmole / hr
Feed temperature	30°C

Table 2. Effect of changing the mass of catalyst on the bottom composition, three reactive stages, ETBE fed in stage 2

Component	Mcat (g per stage)				
	60	70	80	90	100
ETOH	0.0019	0.0019	0.0019	0.0019	0.0001
IB	0.0001	0.0001	0.0001	0.0001	0.001
ETBE	0.9969	0.9969	0.9968	0.997	0.9969
n-B	0.0011	0.0011	0.0012	0.001	0.002

Table 3. Effect of changing the mass of catalyst on the bottom composition, three reactive stages, ETBE and ETOH fed in stage 2

Component	Mcat (g per stage)				
	60	70	80	90	100
ETOH	0.0019	0.0019	0.0019	0.0019	0.002
IB	0.0001	0.0001	0.0001	0.0002	0.0001
ETBE	0.9968	0.9968	0.9967	0.9969	0.9968
n-B	0.0011	0.0012	0.0012	0.001	0.0011

Table 4. Effect of changing the mass of catalyst on the bottom composition, four reactive stages, ETBE fed in stage2

Components	Mcat (g per stage)		
	180	185	190
Ethanol	0.018	0.017	0.0159
IB	0.0015	0.0014	0.0013
ETBE	0.968	0.9699	0.9719
n-B	0.0126	0.0118	0.011

Table 5. Effect of changing the mass of catalyst on the bottom composition, four reactive stages, ETBE and ETOH fed in stage 2

Components	Mcat (g per stage)		
	180	185	190
ETOH	0.04937	0.04872	0.0483
IB	0.002325	0.00222	0.002157
ETBE	0.9455	0.9463	0.9467
n-B	0.0081	0.00774	0.00749

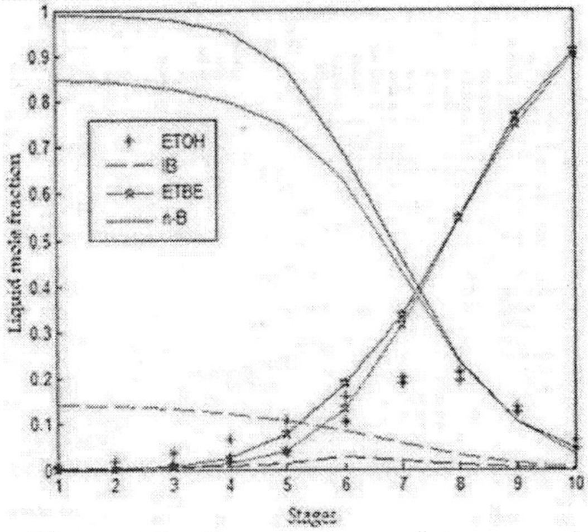

Figure 3. Comparison of proposed program and Aspen Plus

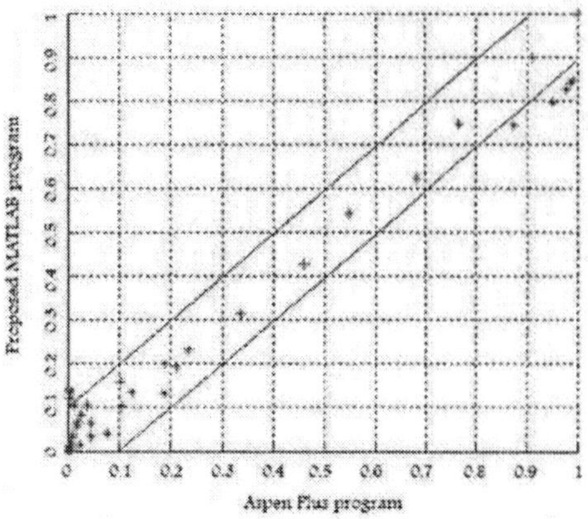

Figure 4. Error of proposed program and Aspen Plus

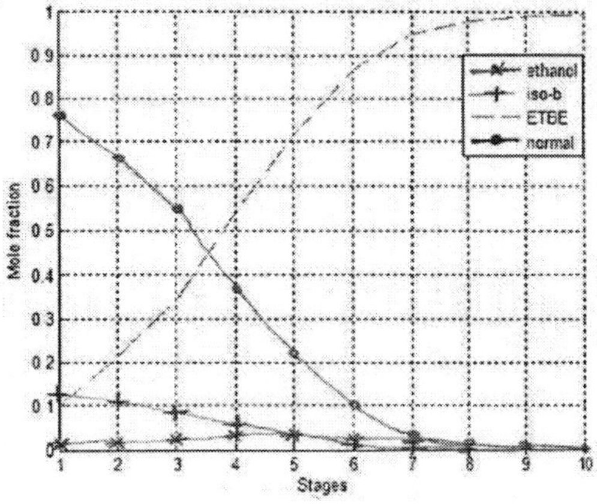

Figure 5. Liquid mole fraction profile when ETBE fed in stage 2

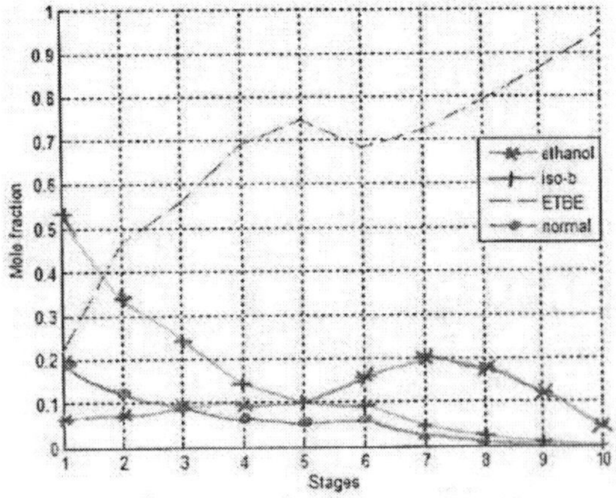

Figure 6. Liquid mole fraction for ETBE fed in stage 2 with four reactive stages

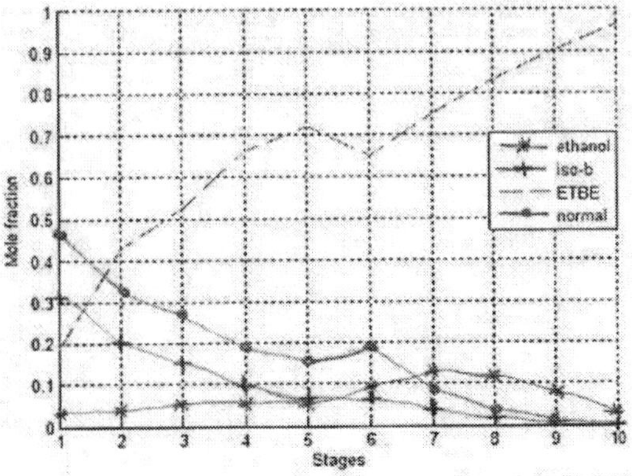

Figure 7. Mole fraction for ETBE & ETOH fed in stage 2 with four reactive stages

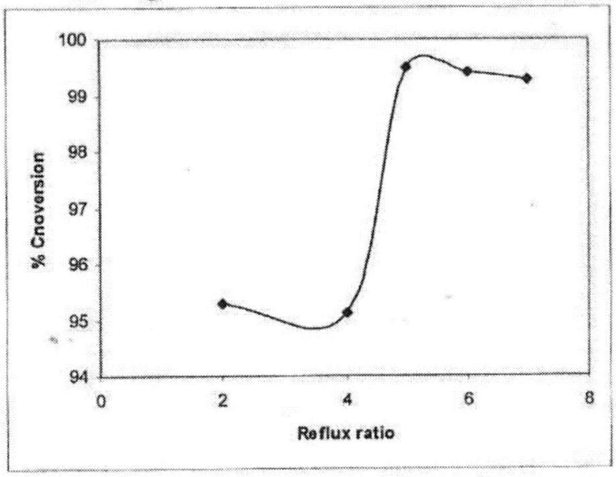

Figure 8. Effect of changing reflux ratio on conversion, 3 reactive stages with ETBE fed to stage 2

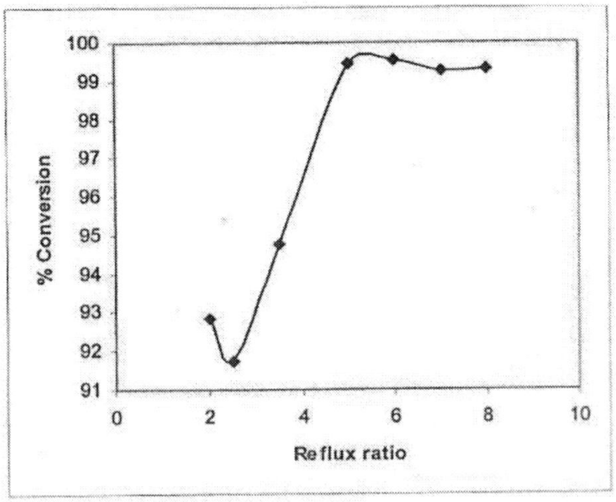

Figure 9. Effect of changing reflux ratio on conversion, 3 reactive stages with ETBE & ETOH fed to stage 2

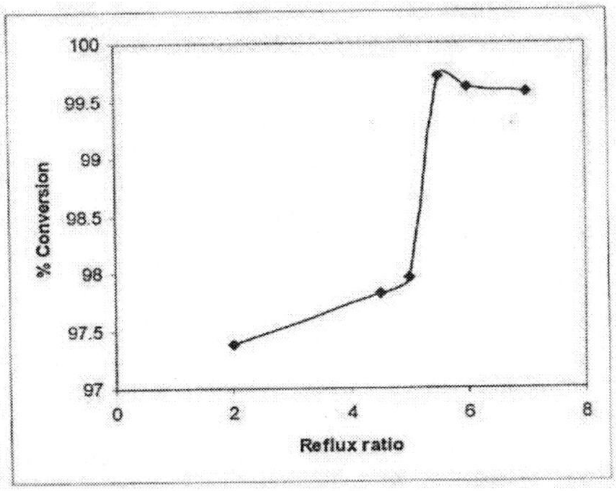

Figure 10. Effect of changing reflux ratio on conversion, 4 reactive stages with ETBE fed to stage 2

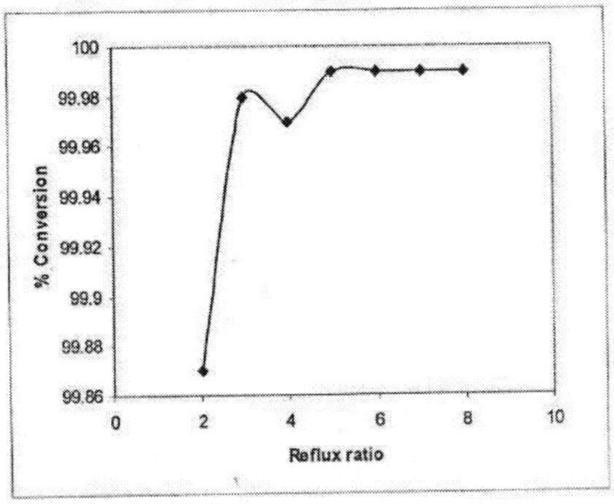

Figure 11. Effect of changing reflux ratio on conversion, 4 reactive stages with ETBE & ETOH fed to stage 2

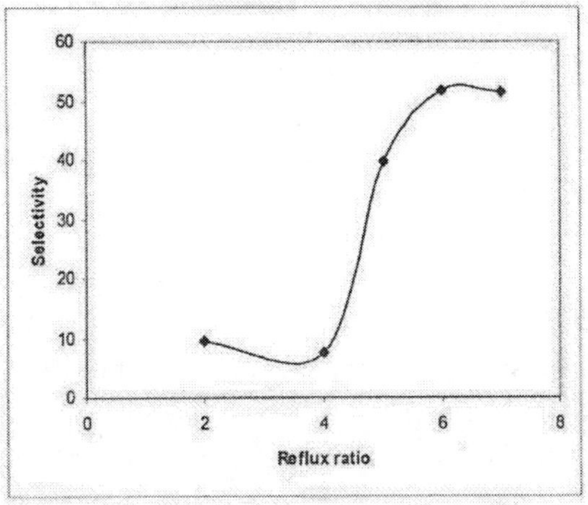

Figure 12. Effect of changing reflux ratio on selectivity, 3 reactive stages with ETBE fed to stage 2

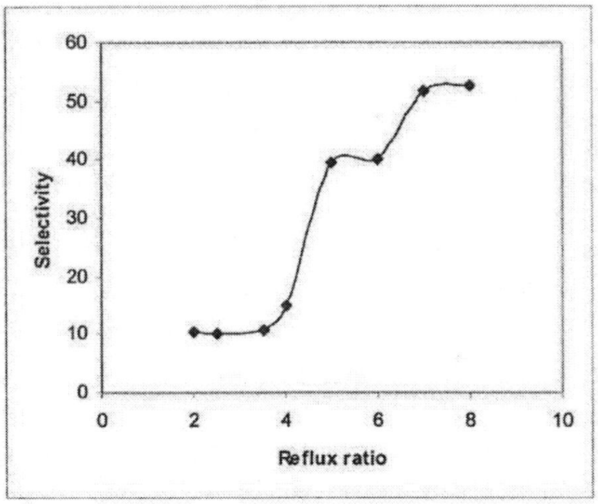

Figure 13. Effect of changing reflux ratio on selectivity, 3 reactive stages with ETBE & ETOH fed to stage 2

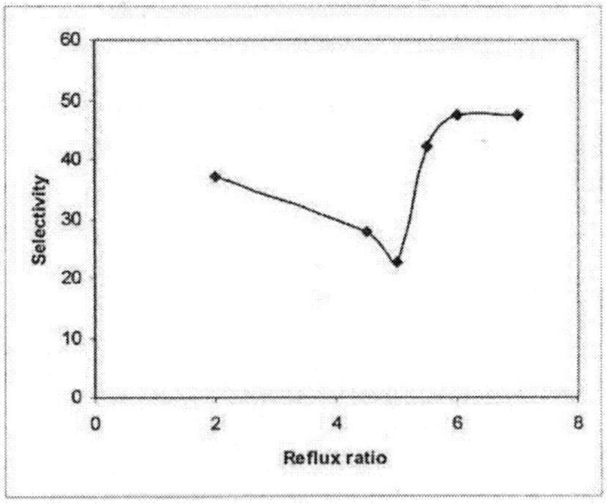

Figure 14. Effect of changing reflux ratio on selectivity, 4 reactive stages with ETBE fed to stage 2

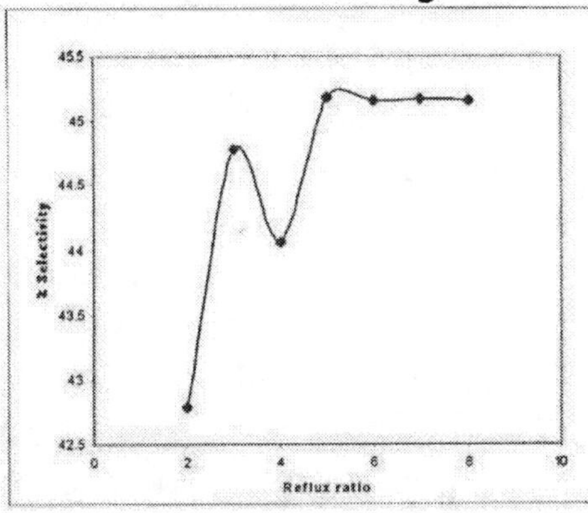

Figure 15. Effect of changing reflux ratio on selectivity, 4 reactive stages with ETBE & ETOH fed to stage 2

Study on mass-transfer performance of a new novel Higee——Rotating Jet Bed

Li Yuming, Wang ying, Ji Jianbing, Wang Guangquan
(College of Chemical Engineering and Materials Science, Zhejiang University of Technology, Hangzhou 310014, China)

Abstract: The Rotating Jet Bed(RJB) has been developed by Zhejiang University of Technology, as a more novel type of gas-liquid mass-transfer equipment than Rotating Zigzag Bed(RZB). The rotor of RJB is composed of perforated concentric baffles fixed to a rotating disc. The top of the baffles is in the grooves of the stationary disc, preventing short cut of the gas. Experiments of total reflux distillation with three types of Higee were carried out with ethanol-water systems at atmospheric pressure. The three types of Higee are RJB with rotor of mesh baffles, RJB with rotor of shutter baffles and RZB. The results showed that the number of theoretical plates per meter of RJB, lower than that of RZB, increased with the increase of rotational speed, and increased firstly and then decreased with the increase of gas superficial kinetic energy F-factor. The maximum number of theoretical plates per meter is 40.7 for RJB with rotor of mesh baffles and 40.0 for RJB with rotor of shutter baffles. The specific pressure drop of RJB was far less than that of RZB. When the gas kinetic energy F-factor was 1.1 $m.s^{-1}(kg.m^{-3})^{0.5}$, the ratios of the specific pressure drop of RJB with rotor of mesh baffles and shutter baffles to that of RZB were 10.0% and 13.6%, respectively. The gas-phase film volumetric mass-transfer coefficients of RJB with rotor of mess baffles and shutter baffles were correlated with the Reynolds number, Grashof number and Schmidt number. The calculated value agreed well with the experimental value, and the average deviation was less than 15%.

Keywords: distillation, rotating jet bed, mass-transfer, specific pressure drop

1. Introduction

The Rotating Jet Bed(RJB) has been developed by Zhejiang University of Technology, as a more novel type of gas-liquid mass-transfer equipment than Rotating Zigzag Bed(RZB)[1], as showed in Fig. 1. The rotor of the RJB is comprised of a rotating disc and a stationary disc. A series of concentric circular sheets are fixed on the rotating disc, which is called the rotational baffles. The rotational baffles are perforated, rotating with the lower(rotating) disc driven by a motor through a vertical shaft. There exists a series of concentric circular grooves on the stationary disc. The top of the rotational baffles enter into the grooves, no contact with the grooves. The perforated rotational baffles offer the flow channels for gas and liquid phase. And the grooves on the stationary disc form the labyrinth seal, preventing the gas from making a short cut. At the eye of the rotor, a special novel liquid distributor is fixed to the rotating disc, rotating with the rotating disc, which is called the rotating liquid distributor.

The liquid is fed through the rotating liquid distributor as jets or droplets onto the most inside rotational baffle. The liquid attaches and flows over the perforations of rotational baffle radially outwards

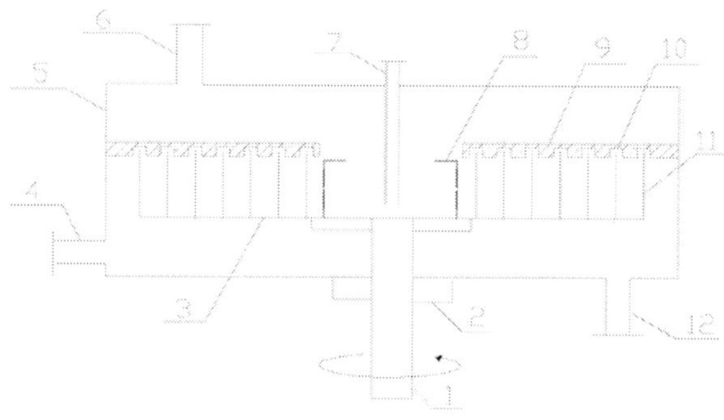

Fig.1 Schematic diagram of rotating jet bed(RJB)

1-vertical rotating shaft; 2-dynamic seal; 3-rotating disc; 4-gas inlet; 5-casing; 6-gas outlet; 7-liquid inlet; 8-rotating liquid distributor; 9-stationary disc; 10-grooves on the stationary disc forming labyrinth seal; 11-rotational baffle; 12-liquid outlet by the centrifugal force as fine jets or droplets. The liquid leaves the most outside rotational baffle, collected by the casing wall and leaving the RJB through the liquid outlet at the bottom of the casing.

The gas is tangentially introduced into the casing, flowing through the perforations of the most outside rotational baffle, and is forced to flow radially inwards countercurrent to the liquid by pressure driving force. The gas leaves the RJB through the gas outlet. The mass transfer takes place in the perforations of the rotational baffles, between two adjacent rotational baffles, and between the rotor and the casing.

The RJB has the advantages of low pressure drop, high flux, high efficiency and low power consumption, overcoming the disadvantages of excess high pressure drop and excess high power consumption of RZB.

In the paper, a total reflux distillation experiment was carried out for the study of the performance of RJB and RZB.

2. Experiment

A total reflux distillation experiment was carried out with ethanol-water system at atmospheric pressure. The performance of the RJB and RZB was studied and the structures of the RJB and RZB were illustrated in Fig.2.

The axial height of the RJB rotor is 47mm. And the RJB rotor is composed of 7 rotational baffles with their diameter of 140mm, 160mm, 185mm, 205mm, 230mm, 250mm, and 270mm. The rotational baffles of RJB employed two type of structure of mesh baffle and shutter baffle, as shown in Fig.3. The mesh baffle is perforated with diamond holes of 0.8mm in side length. The shutter baffle has plenty of shutter holes of 35mm in length, 10mm in width and 20°in opening angle.

The axial height of the RZB rotor is also 47mm. And there are 9 rotational and 10 stationary baffles for the RZB rotor. The rotational and stationary baffles are 43mm and 30mm in height, respectively. The upper part of the rotational baffles is perforated with holes of 1.5mm in diameter.

The experiment process of the total reflux distillation experiment was shown in Fig.4.

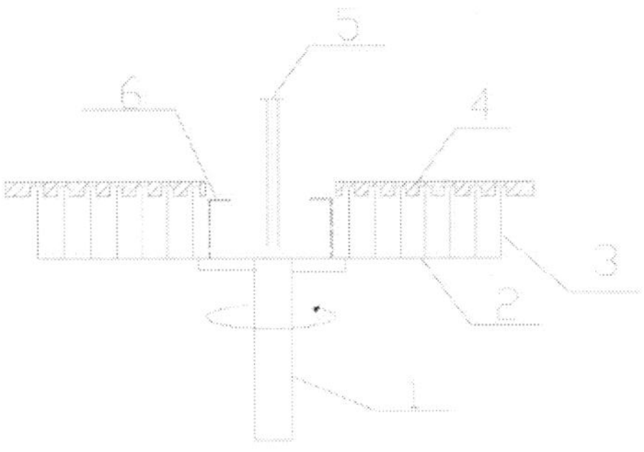

(a) rotor of RJB

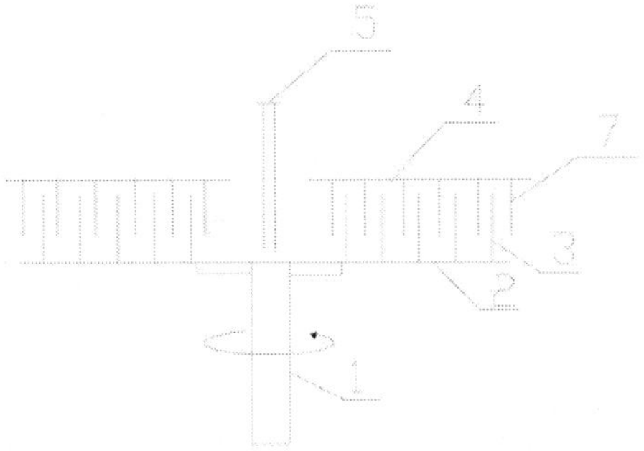

(b)rotor of RZB

Fig.2 Schematic diagram for rotor

1-rotating shaft; 2-rotating disc; 3-rotating baffles; 4-stationary disc; 5-liquid inlet; 6-liquid distributor; 7-stationary baffles

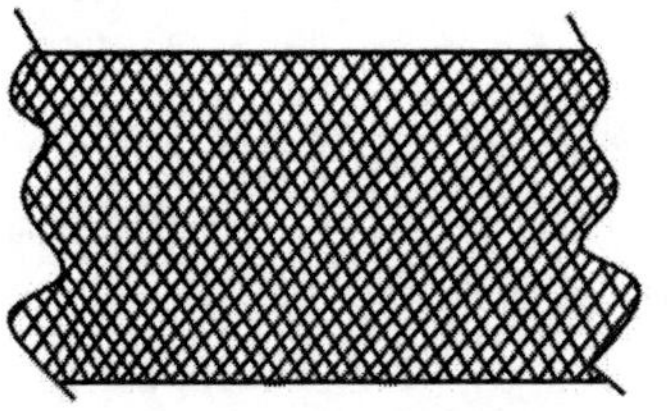

(a) mesh baffle in RJB rotor

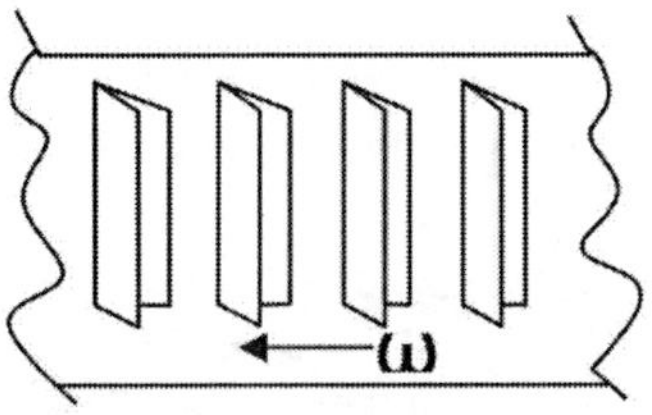

(b) shutter baffle in RJB rotor

Fig.3 Structure of baffles in RJB rotor

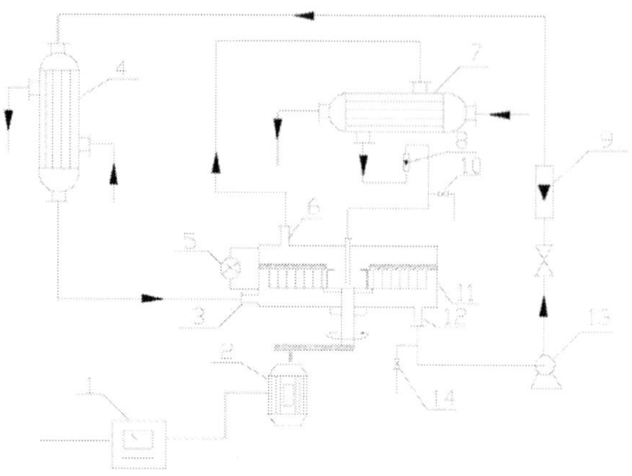

Fig.4 Experimental setup of distillation in RJB/RZB

1-frequency modulator; 2-motor; 3-gas inlet; 4-reboiler; 5-pressure difference sensor; 6-gas outlet; 7-condenser; 8,9-rotameter; 10-top sample; 11-rotating jet bed(RJB); 12-liquid outlet; 13-pump; 14-bottom sample

The liquid was vaporized into the vapor in the reboiler using heat-conducting oil. The vapor flowed into the RJB/RZB through the gas inlet. Leaving the RJB, the vapor is condensed into the liquid in the condenser. The liquid flowed into the RJB/RZB through the liquid inlet, leaving the RJB/RZB and pumped into the reboiler.

The pressure drop between gas inlet and outlet was measured with a pressure difference sensor. The reflux rates of liquid from the condenser were measured with a rotameter. The liquid samples at liquid inlet and outlet was taken to determine the liquid concentration and the number of theoretical plates.

3. Result

3.1 Effect of gas superficial kinetic energy F-factor on the number of theoretical plates per meter

The gas superficial kinetic energy F-factor of RJB/RZB is defined as $Q_G\sqrt{\rho_G}/(2\pi r_1 z)$, where Q_G is vapor rate, m^3/s; ρ_G is vapor density, kg/m^3 ; r_1 is the inner diameter of the rotor, m; and z is the height of the rotor, m.

The number of theoretical plates per meter is defined as $N_T/(r_2 - r_1)$, where N_T is the number of theoretical plates; r_2 is the outer diameter of the rotor, m.

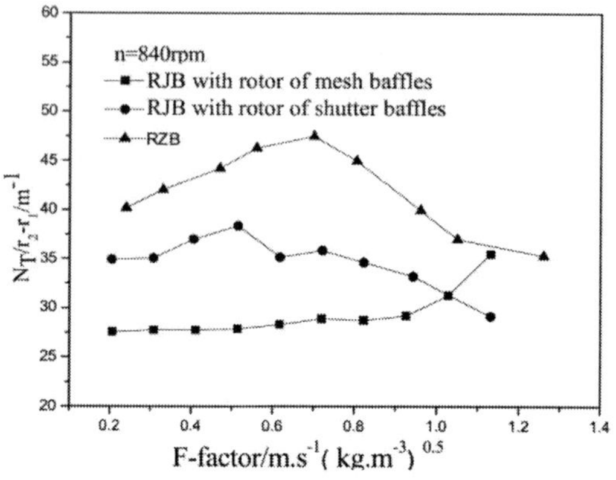

(a) rotational speed of 840rpm

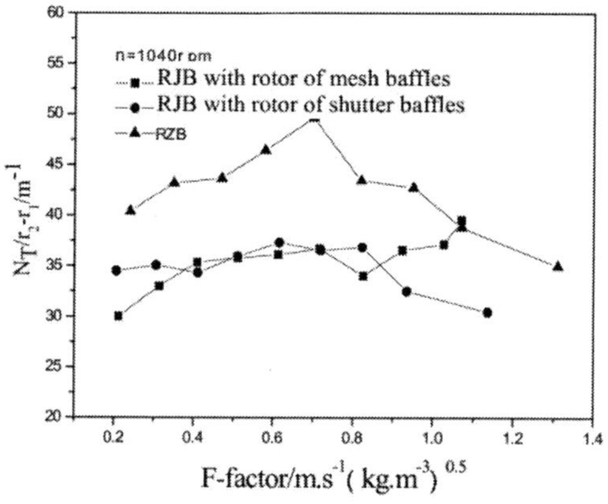

(b) rotational speed of 1040rpm

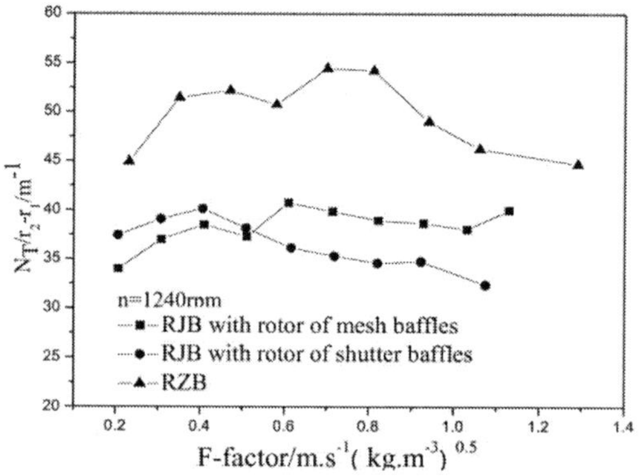

(c) rotational speed of 1240rpm

Fig.5 Effect of gas superficial kinetic energy F-factor on number of theoretical plates per meter

The effect of gas superficial kinetic energy F-factor on the number of theoretical plates per meter was shown in Fig.5.

The number of theoretical plates per meter, $N_T/(r_2-r_1)$, of RJB was lower than that of RZB, occupying 72.3~89.1% of that of RZB. However, $N_T/(r_2-r_1)$ of RJB still reached 30~40; and the maximum $N_T/(r_2-r_1)$ is 40.7 for RJB with rotor of mesh baffles and 40.0 for RJB with rotor of shutter baffles.

The number of theoretical plates per meter, $N_T/(r_2-r_1)$, of RJB increased firstly and then decreased with the increase of gas superficial kinetic energy F-factor; and $N_T/(r_2-r_1)$, of RJB increased with the increase of rotational speed.

3.2 Effect of gas superficial kinetic energy F-factor on the specific pressure

The specific pressure is defined as $\Delta P/N_T$,

where P is the pressure drop of RJB/RZB, Pa.

The effect of gas superficial kinetic energy F-factor on the specific pressure were shown in Fig.6.

The specific pressure drop, $\Delta P/N_T$, of RJB was far less than that of RZB. When the gas kinetic energy F-factor was 1.1 $m.s^{-1}(kg.m^{-3})^{0.5}$, the ratios of $\Delta P/N_T$ of RJB with rotor of mesh baffles and shutter baffles to that of RZB were only 10.0% and 13.6%, respectively.

The specific pressure drop, $\Delta P/N_T$, of RJB increased with the increase of gas superficial kinetic energy F-factor and rotational speed.

3.3 The gas-phase volumetric mass-transfer coefficients of RJB

For the packed tower, the relation of the number of theoretical plates per meter, N_T/H, and the height of a transfer unit(HTU), H_{OG}, is

$$\frac{N_T}{H}=\frac{1}{H_{OG}}\cdot\frac{\lambda-1}{\ln\lambda} \quad (1)$$

For the rotating bed, modifying the eq(1) is

$$\frac{N_T}{r_2-r_1}=\frac{1}{H_{OG}}\cdot\frac{\lambda-1}{\ln\lambda} \quad (2)$$

The height of a transfer unit(HTU), H_{OG}, multiplied by the number of transfer units (NTU), N_{OG}, is

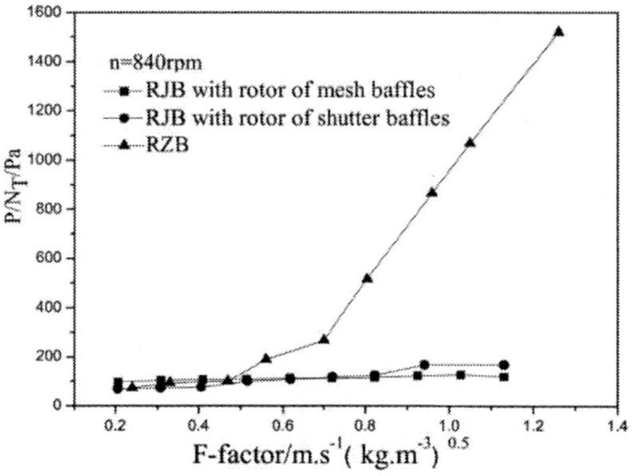

(a) rotational speed of 840rpm

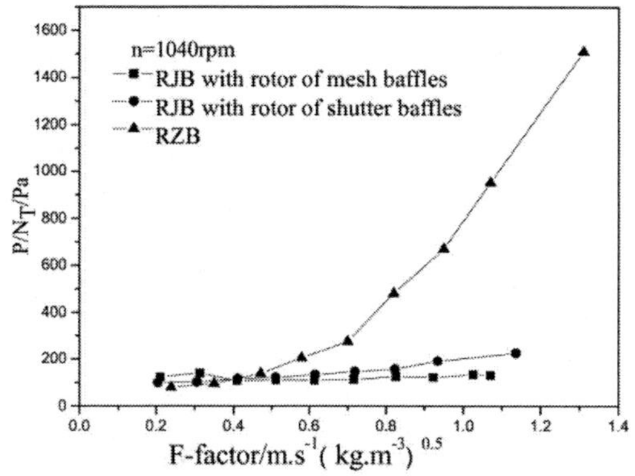

(b) rotational speed of 1040rpm

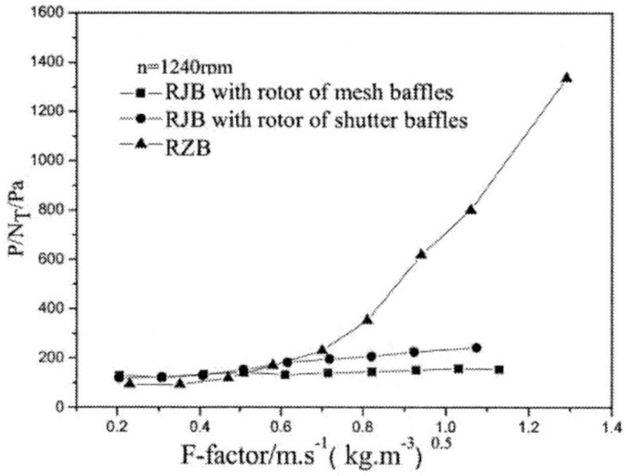

(c) rotational speed of 1240rpm

Fig.6 Effect of gas superficial kinetic energy F-factor on specific pressure drop

$$r_2 - r_1 = H_{OG} \cdot N_{OG} \quad (3)$$

Substituting eq(2) into eq(3) is

$$N_{OG} = \frac{r_2 - r_1}{H_{OG}} = N_T \cdot \frac{\ln \lambda}{\lambda - 1} \quad (4)$$

where $\lambda = m \cdot (V / L)$, and m is average slope of vapor-liquid equilibrium curve, V/L is the ratio of vapor to liquid.

The gas overall volumetric mass transfer coefficient is [2]

$$K_G a_e = \frac{Q_G}{z\pi(r_2^2 - r_1^2)} \cdot N_{OG} \quad (5)$$

Substituting eq(4) into eq(5) is

$$K_G a_e = \frac{Q_G}{z\pi(r_2^2 - r_1^2)} \cdot N_T \cdot \frac{\ln \lambda}{\lambda - 1} \quad (6)$$

From eq(6), the gas overall volumetric mass transfer coefficient may be obtained.

Due to the gas-film control in the distillation, the gas overall volumetric mass transfer coefficient, $K_G a_e$, is equal to the gas film volumetric mass transfer coefficient, $k_G a_e$. The correlation of $k_G a_e$ is [3]

$$k_G a_e = C(a_p D_G / d_p) Re^x Gr^y Sc^z \quad (7)$$

where Reynolds number is $Re = \frac{G_G}{a_p \mu}$, Schmidt number is $Sc = \frac{\mu}{\rho D}$, Grashof number is $Gr = \frac{d_p^3 \rho^2 r \omega^2}{\mu^2}$; a_p is surface area of $216 m^2/m^3$ for RJB with rotor of mesh baffles and $256 m^2/m^3$ for RJB with rotor of shutter baffles; d_p is hydraulic diameter of 0.77mm for RJB with rotor of mesh baffles and 15.5mm for RJB with rotor of shutter baffles; and the exponent, z, of Schmidt number is assumed to -1/3[7].

Using the experimental data, The correlations of $k_G a_e$ of RJB with rotor of mesh baffles and shutter baffles are as follows:

RJB with rotor of mesh baffles:

$$k_G a_e = 10.327(5.077) Re^{1.13} Gr^{0.30} Sc^{-1/3}$$

RJB with rotor of shutter baffles:

$$k_G a_e = 128.98(0.419) Re^{0.935} Gr^{0.053} Sc^{-1/3}$$

The calculated value agreed with the experimental value, with the greatest error of 14.05% and average error of 4.98% for RJB with rotor of mesh baffle; and with the greatest error of 9.58% and average error of 4.07% for RJB with rotor of shutter baffle.

4. Conclusion

RJB is superior to RZB. Although the number of theoretical plates per meter of RJB is slightly lower than that of RZB, the specific pressure drop of RJB was far less than that of RZB.

The number of theoretical plates per meter of RJB reached 30~40; and the specific pressure drop of RJB is only 200Pa.

The gas-phase film volumetric mass-transfer coefficients of RJB were correlated with the Reynolds number, Grashof number and Schmidt number.

References

[1] G.Q. Wang, Z.C. Xu, J.B. Ji, Performance of a rotating zigzag bed——A new HIGEE, Chemical Engineering and Processing: Process Intensification, v 47, n 12, p 2131-2139, **2008**

[2] Yu-Shao Chen, Chia-Chang Lin, Mass transfer in a rotating packed bed with viscous newtonian and non-Newtonian fluids, v 44, n 4, p 1043-1051, **2005**

[3] Kelleher T , Fair J R Distillation studies in a high-gravity contactor[J] Industry & Engineering Chemical Reseach **1996** , 35 4646-4655.

AUTHOR INDEX